Writing History

General Editors: Stefan Berger
                  Heiko Feldner
                  Kevin Passmore

Also in the Writing History series

Published:

Writing History: Theory and Practice
Edited by Stefan Berger, Heiko Feldner and Kevin Passmore

Writing Gender History
Laura Lee Downs

Writing Medieval History
Edited by Nancy Partner

Forthcoming:

Writing Early Modern History
Edited by Garthine Walker

The Holocaust and History
Wulf Kansteiner

Writing Medieval History

Edited by: Nancy Partner

Hodder Arnold

A MEMBER OF THE HODDER HEADLINE GROUP

First published in Great Britain in 2005 by
Hodder Education, a member of the Hodder Headline Group,
338 Euston Road, London NW1 3BH

http://www.hoddereducation.co.uk

Distributed in the United States of America by
Oxford University Press Inc.
198 Madison Avenue, New York, NY10016

The advice and information in this book are believed to be true and
accurate at the date of going to press, but neither the author nor the publisher
can accept any legal responsibility or liability for any errors or omissions.

British Library Cataloguing in Publication Data
A catalogue record for this book is available from the British Library

Library of Congress Cataloging-in-Publication Data
A catalog record for this book is available from the Library of Congress

ISBN 0 340 80845 4 (hb)
ISBN 0 340 80846 2 (pb)

1 2 3 4 5 6 7 8 9 10

Typeset in 11pt Adobe Garamond by Servis Filmsetting Ltd, Manchester
Printed and bound in Great Britain by CPI Bath

What do you think about this book? Or any other Hodder Education title?
Please send your comments to the feedback section on www.hoddereducation.co.uk

Contents

Notes on contributors

Cordelia Beattie is Lecturer in Medieval History at the University of Edinburgh. She has published various articles on women in late medieval England. Her research interests include gender and work, and women's experience of the law, particularly in the English court of Chancery, and she is currently completing a monograph on 'Medieval Single Women: Categorizing Women in England *c*.1275–1525'.

Sarah Foot is Professor of Early Medieval History at the University of Sheffield. She has published extensively on the early English Church, including *Veiled Women* (2000) and *Minsters: Monastic life in England* c.*600–950* (CUP, forthcoming). Her other research interests include the development of English identity before the Conquest and the ways in which the Anglo-Saxons constructed usable pasts. She is currently co-editing the Anglo-Saxon charters of Bury St Edmunds Abbey for the British Academy and is Director of the 'Cistercians in Yorkshire Project' at the University of Sheffield.

Jacqueline Murray is Professor of History and Dean of the College of Arts at the University of Guelph. Her publications include *Conflicted Identities and Multiple Masculinities: Men in the Medieval West* (1999) and *Love, Marriage and Family in the Middle Ages. A Reader* (2001). Her research interests include the history of women and sexuality and gender in medieval society. She is currently working on a study of the understanding of masculinity and male sexuality in the Middle Ages.

Derek Neal recently completed his PhD in History at McGill University. He has published 'Suits Make the Man: Masculinity in Two English Law Courts, *c.* 1500', *Canadian Journal of History* 37 (April 2002), pp. 1–22. His article 'Husbands and Priests: Masculinity, Sexuality and Defamation in Late Medieval England', in *The Hands of the Tongue: Essays on Deviant Speech*, edited by Edwin Craun, is currently in press with Medieval Institute Publications of Kalamazoo, Michigan. A book, *False Thieves and True Men: Meanings of Masculinity in Late Medieval England*, is currently being submitted to publishers. His next project is a study of masculinity among the English clergy, 1460–1560.

Monika Otter is Associate Professor of English and Comparative Literature at Dartmouth College. Her publications include 'Inventiones: Fiction and Referentiality in Twelfth-Century English Historical Writing' (Chapel Hill: North Carolina University Press, 1996), a translation of Goscelin of St Bertin's 'Liber Confortatorius' (Boydell and Brewer, 2004) and articles on medieval historiography and medieval Latin literature. She is currently working on a study of the use of the first person and notions of literary character in the central Middle Ages.

Nancy Partner is Professor of History at McGill University. She has published numerous articles on historiography and historical theory, including 'Medieval History and Modern Realism: Yet Another Origin of the Novel', *MLN* (1999), 'Hayden White: The Form of the Content', *History and Theory* (1998), and 'Hayden White (and the Content and the Form and Everyone Else) at the AHA', *History and Theory* (1998). She edited the book, *Studying Medieval Women: Sex, Gender, Feminism* (Cambridge, MA, 1993). Her research interests include narrative theory and historical writing, and psychoanalysis in relation to history.

Jay Rubenstein is Assistant Professor of History at the University of New Mexico. His publications include the book, *Guibert of Nogent: Portrait of a Medieval Mind*, as well as several articles touching on intellectual history, biblical exegesis, the crusade movement, and the historians and historiography of the Norman Conquest. His next project will examine how contemporaries understood and interpreted the historical meaning of the capture of Jerusalem in 1099.

David Gary Shaw is Professor and Chair of History at Wesleyan University, and Associate Editor of *History and Theory*. His publications include *The Creation of a Community. The City of Wells in the Middle Ages* (1993), *The Return of Science: Evolution, History and Theory*, edited with Philip Pomper (2002), and *Necessary Conjunctions: The Social Self in the Middle Ages* (forthcoming).

Robert M. Stein is Associate Professor of Language and Literature at Purchase College, SUNY, and Adjunct Professor of English and Comparative Literature at

Columbia University. He has published articles on modern critical theory as well as on medieval romance and historiography, and is the editor, with Sandra Pierson Prior, of *Reading Medieval Culture* (University of Notre Dame Press, forthcoming). He is currently completing a book-length study of twelfth-century romance and historiography.

General editors' preface

Can historical writing tell us anything about the past given that many – post-structuralists in the lead – would deny that academic historical writing is intrinsically different from fiction? Does the study of the past serve any purpose in a society in which, according to Eric Hobsbawm, 'most young men and women . . . grow up in a sort of permanent present lacking any organic relations to the public past of the times in which they live'?

Historians have never been more inclined to reflect upon the nature of their discipline. Undergraduate and postgraduate courses increasingly include compulsory study of historiography, the philosophy of history and history and theory. *Writing History* presents a book series which focuses on the practical application of theory in historical writing. It publishes accessibly written overviews of particular fields of history. Rather than focus upon abstract theory, the books in this series explain key concepts and demonstrate the ways in which they have informed practical work. Theoretical perspectives, acknowledged and unacknowledged, have shaped actual works of history. Each book in the series relates historical texts and their producers to the social conditions of their existence. As such, *Writing History* does go beyond a focus on historical works in themselves. In a variety of ways, each volume analyses texts within their institutional arrangement and as part of a wider social discourse.

Nancy Partner's *Writing Medieval History* explores a wide variety of popular theoretical approaches to the writing of medieval history over the past two decades. The times when a structural social history held sway over the study of the Middle Ages, either in the form of Marc Bloch's *Feudal Society* or Otto Hintze's and Otto Brunner's studies, have long gone. Instead, with the revival of narrative since the 1980s, historians have been increasingly interested in exploring the diverse ways in which forms of subjectivities have been constructed.

Natalie Zeman Davis's *The Return of Martin Guerre* (1983) was perhaps one of the most visible expressions of this new-found interest in the medieval *dramatis personae*. How were they similar to the subjectivities constructed in the contemporary world, and what made them different? The first three chapters of the book review the responses of medieval historians to this question. They deal specifically with the self in the Middle Ages, highlighting the impact of psychoanalytic theory, particularly on the study of biographical and autobiographical texts, and on the field of medieval history more generally.

As theories of narrativity gained ground within the historical profession, the influence of literary criticism has left its mark on historical writing. The next three chapters discuss this influence on medieval historians who have shown far more interest in studying fictional texts of the past than have modernists. They have also become more versed in analysing narrative techniques of specific historical texts.

Finally, no other approach to history has challenged both traditional understandings of history as a science and established grand narratives to the extent that gender history has. The use of gender as a crucial category of historical analysis has permitted historians to open entirely new areas of study. This development has also left its mark on the field of medieval history, as the last set of essays powerfully reminds us. Overall, the reader is presented with a multitude of intriguing insights on how gender studies, postmodernism, the 'new historicism', psychoanalysis and theories of identity have reconfigured the field of medieval history in the past two decades.

Stefan Berger, Heiko Feldner and Kevin Passmore
Cardiff and Pontypridd, June 2004

Preface

The post-traditional Middle Ages: the distant past through contemporary eyes

When Sir Richard Southern (surely one of the most brilliant medievalists of the twentieth century) delivered his inaugural lecture upon becoming the Chichele Professor of Modern History at the University of Oxford in 1961, he was entering an eminent position in a field that had established a grip on its subject matter, methods and interpretive ideas only a very few generations earlier.

'Medieval history' in our academic sense of the term, had been a radical innovation in university life when the very first Oxford professorship of Modern (that is, post-Roman) History was created in 1862. A mere blink of historical time makes medieval history a far more recent entry among the post-classical subjects of systematic study than most people would guess, and also oddly makes medieval history the oldest of the officially 'modern' historical fields. Southern's inaugural address – reprinted as 'The Shape and Substance of Academic History', in *History and Historians: Selected Papers of R. W. Southern* – ought to be required reading (a required pleasure) for all students of history, for many reasons starting with the author's sharp prescience and humour about the early efforts of his discipline to find its intellectual footing as a field of academic research and teaching.[1] After decades (the 1840s and 1850s) embroiled in unresolvable theological controversy, the University of Oxford was to be restored to vitality, to national purpose and modernity – by history, the proper education 'for gentlemen, for men of affairs, for open-minded men, free from the cobwebs of useless learning and ancient error'.[2]

'There remained, however, one great difficulty. It was all very well to establish modern history as an academic discipline; but the question still remained, how was it to be studied?'[3] Or even *what* was to be studied. As Southern drily and funnily explains, no one had a clue as to how to study history, or what history to study, once academics ventured past the events contained in the texts of a few canonical authors of Greek and Latin antiquity – Thucydides, Sallust, Tacitus, and others already part of the classics curriculum. Even for these historian authors, study involved philological commentary on the text, with an uncritical assumption that the ancient authorities were to be read in a spirit of scrupulous reverence because they were without error or bias in their information. This new field of modern (we must remember this term essentially meant after the ancients) history was frustratingly devoid of authoritative texts like those of the ancients, and thus threatened to be a historical field empty of history, with nothing at all in it to study, much less a viable method for studying it. The well-established techniques of philology were based in classical Greek and Latin and there seemed to be no authoritative historians of the post-classical centuries for the professors of modern history to lecture and comment on. History, in its first several years at Oxford, shambled along as a slack, confused subject, openly regarded as an easy time for unambitious gentlemen. 'The truth was that history had attained academic status in 1850 on a wave of opposition to theological dogmatism and impatience with ancient restrictions, without anyone being clear whether the subject had a method, or a public, or indeed whether it was a recognizable subject at all.'[4]

Academic discipline emerged in England with William Stubbs (Regius Professor from 1867 to 1884) and the systematic study of constitutional history: 'Intellectually it was highly respectable. It was systematic; it gave an organic unity to a large assortment of otherwise disconnected events. It was difficult.'[5] History at the English universities did not match the fervour of continental scholarship – especially the German school led by Ranke – for rigour, scientific aspirations or scholarly publication. The seminar system, which originated in Germany, for intense group engagement with primary sources, was transplanted direct to US universities, while young Englishmen of the late nineteenth century were prepared by a more practical if less systematic history to become the generalist statesmen and administrators of the Empire.[6] To add a rather obvious comment to Southern's nuanced analysis of this state of affairs, the study of English constitutional history, with its (then) self-evident lines of development, dismissed now by the term 'whig history', hardly invited or seemed to need special methodology, much less self-conscious theory.

'Nothing', he notes, 'is more striking about the Oxford historians of the late nineteenth century than their disregard of everything in history that could not be related to institutions and politics. They left out that which is most interesting in the past to concentrate on that which was practically and academically most serviceable.'[7] Those 'most interesting' parts of historical life which Southern notes as most completely ignored and absent from history as it developed at Oxford, and, I would add, elsewhere in the English-speaking world, were exactly what was emerging strongly in the 1960s. (Southern modestly does not mention his own contributions to a subtle and deeply humane approach to medieval life.) The leading motif of Southern's analysis of where his field was, in 1961, and where it was tending, was 'enlargement' – expansion into all the areas previously ignored and many that had never been considered part of historical study at all. His inaugural lecture, a trenchant yet unpolemical and gracious account of why the extended moment of unquestioned coherence in post-Roman history had to end, along with the illusory stasis of its 'world which disappeared for ever in 1914', envisions history opening into unpredictable and probably fragmented connections with other disciplines. And he was right, of course.

Southern's own, and most personally characteristic, definition of history's compelling subjects is: 'the study of the thoughts and visions, moods and emotions and devotions of articulate people. These are the valuable deposit of the past.'[8] The topics and approaches in the essays in this volume of the *Writing History* series are, as I see it, perfectly continuous with this insight, and with Southern's acknowledgement that 'the greatest developments in historical thought have been on the periphery of the old syllabus' while what used to be the centre was 'quietly ceasing to be the centre'. He looked ahead to 'surveys of problems or of tracts of history which have at present no place in our syllabus'.[9]

Social history, in the second half of the twentieth century, moved the periphery to the centre. The great 'tract' of human experience unrecognized by the traditional syllabus, that of people largely excluded from the centres of political life and institutions of governance, opened the way towards later and continuing investigations of private life, domestic households, the historical situation of women, children, gender and sexuality, attention to culture in the non-elite sense, the inclusion of literature as part of the historical realm. Constitutional history had rested on information from the most readily available documents of medieval history – records of government and the major historians – which yielded their factual information to a critical empirical scrutiny, which felt, to its practitioners, like systematic common sense. Unlike the unself-conscious certainties of constitutional history, whose facts obligingly appeared to fall into patterns of proto-parliamentary (or republican, depending on which side of the Atlantic one was) development, social history brought methodology into self-conscious view. Tracing the history of ordinary people, in social groups and collectivities mostly below the horizon of intentional documentation, demanded new approaches to the archives and openly acknowledged intentions on the part of historians – notably quantitative methods, new scrutinies of legal and economic evidence and marxist-derived conceptions of causation and social structure. Theory, in the core sense of openly conceptualized ideas about society and agency, and about the nature of historical evidence, entered medieval history with social history. When historians enlarged their interests yet further to areas like women's history, gender history and culture in the anthropological acceptance, self-conscious theory tracked the new fields in parallel and ever more sophisticated metahistorical commentary.

The varieties of theory discussed by the historians who contributed to this book are those which emerged from the so-called linguistic turn. These modes of theory sparked acrimony and debate during the past twenty years or so, and have sorted themselves out from their exciting beginnings to join the armature of inter-pretive instruments now indispensable to the contemporary discipline. To varying degrees, each section of the book addresses concepts and techniques all medieval-ists need to understand historical documents as texts as well as sources, construc-tions of language layered upon language as well as respositories of facts. The buried assumptions of positivist history – that the language of evidentiary sources is transparent on its underlying reality, that comparison and context are sufficient to enable the historian to select the 'true facts' from amidst errors in sources, that all usable information is conveyed from one fully conscious mind to another – were challenged wholesale by the discourse analysis, deconstruction, semiotics and narrative studies associated with the linguistic turn. The era of combative rejection and overstated defence (generally the 1980s into the mid-1990s) has exhausted its acrimony. From seeming to many an attack on the historical enter-prise itself, all the critical techniques related to the textuality of sources have quietly taken their place in advanced historical curricula, in graduate training and in research and writing throughout the discipline.

The three parts of this book address and incorporate approaches to the 'read-ability' of the past in ways that are various but coherent. The issue of who exactly we think inhabited the medieval past was easily submerged by both the institu-tional generalities of constitutional history, and the collective interests of social history. Contemporary interest in restoring the voice and agency of ordinary people, the poor, the young, most especially women, brings the self into focus as an object of inquiry. Medieval people lived much of their lives in public view, open to critical comment from family, employers and neighbours, and thus were exposed fully to the pressures for social and moral conformity. They typically expressed their individuality, deepest wishes and resistance to authority in lan-guage or behaviour that is superficially conventional and yet leaves indubitable traces of the individual mind. Part 1, 'Recognizing people in medieval society: the self', acknowledges in chapter 1 the crucial social interface of medieval identity, as honour, reputation and fullness of personal presence were daily negotiated in small-scale public interchanges. Chapter 2, moving from social behaviour to lit-erary expression, unpacks the wealth of latent authorial self-revelation in every form of narrative. Chapter 3 advances the case for psychoanalytic theory as the key conceptual framework for recognizing the interior self and unconscious wishes in medieval people.

Medieval texts that approached their contemporary readers, and us, claiming to be non-fictional works of history nevertheless drew fully on the paradigms of con-temporary fiction. They made use of the same techniques of narrative structure, were conscious of their relations with other texts and played with the poetic resources of the language in ways that we associate with works of fiction. In this section of essays, Part 2, 'Literary techniques for reading historical texts', chapter 4 offers the appropriate critical language and reading techniques for approaching historical evidence in both its documentary function and textuality. Chapter 5 extends narrative theory to expose the previously unrecognized formal structure of annals and chronicles. Chapter 6 discusses the formal properties and rhetorical functions of fiction inside medieval historical writing.

Sexuality and gender have entered the range of medieval historical topics because historians have recognized that people inevitably think about sex and sexual identities with the conceptual language available in their cultural surroundings. The human body and its reproductive biology may have been the same in the thirteenth century as it is now, but ideas about the body, sexual activity and the differences between the sexes were formed by a historically specific language, grounded in Christian morality and ancient medical authority. Discourse analysis is the key instrument historians now use to separate out culturally constructed formulations about human sexuality and expose the frequent ambiguities and self-contradictions inherent in the language of 'nature' and 'natural' behaviour. Using similar analytic techniques, historians examine medieval conceptions of gender, the specific strengths, weaknesses, virtues and vices presumed to be characteristic of females (femininity) and males (masculinity), opening new perspectives on the foundations of social hierarchy and economic and legal inequalities. The essays of Part 3, 'Historicizing sex and gender', foreground the analytical techniques and conceptual language which have given human sexuality and gender a history. Chapter 7 demonstrates exactly why sex in the Middle Ages has to be localized in social, religious and cultural terms; chapter 8 displays the variability and self-contradictions inherent in medieval constructions of the feminine; and chapter 9 opens the new masculinity studies, based on the premise that men and masculinity are not locked together in any single idealized or permanent form.

The purpose of this book is to collect together the core varieties of the 'new theory' with some of their applications to research which have proved their usefulness beyond doubt for medieval history. These theory-informed approaches are here to stay in the discipline. They have become necessary knowledge and skills for the formation of any medieval historian – major fields of research are impossible without them. Informed readers will no doubt notice, and variously regret, some absences. What is not here was given considerable thought. There are some quite interesting contenders for inclusion that, at length, I decided did not meet the criterion of here-to-stay proven usefulness, or not yet. These included medieval postcolonialism, queer theory and cultural studies. Another later book along these same lines will not look the same as this, and it should not. Sources will never settle back into seeming transparent passive containers of good and dubious facts; medieval people will always now decline to be obedient idiographs of a didactic cultural ideal; sex has won its right to have a history and gender – feminine or masculine, will never again appear as a specific character type inevitably attached to a sexed body.

Nancy Partner
July 2004

Notes

1   Richard Southern, 'The Shape and Substance of Academic History', in Richard Southern and Robert Bartlett, *History and Historians: Selected Papers of R.W. Southern* (Oxford, 2004), ch. 5.
2   Southern, 'The Shape and Substance of Academic History', p. 91.
3   Southern, 'The Shape and Substance of Academic History', p. 91.
4   Southern, 'The Shape and Substance of Academic History', p. 92.
5   Southern, 'The Shape and Substance of Academic History', p. 94.
6   Southern, 'The Shape and Substance of Academic History', p. 96. For the not quite parallel history of medieval studies in the USA, in which the European and, especially, British Middle Ages moved from republican origins to radical alterity, see Paul Freedman and Gabrielle M. Spiegel, 'Medievalisms Old and New: The Rediscovery of Alterity in North American Medieval Studies', *American Historical Review* 103 (1998), pp. 677–704.
7   Southern, 'The Shape and Substance of Academic History', p. 99.
8   Southern, 'The Shape and Substance of Academic History', p. 100.
9   Southern, 'The Shape and Substance of Academic History', p. 101.

Part 1

Recognizing people in medieval society: the self

1

Social selves in medieval England: the worshipful Ferrour and Kempe

David Gary Shaw

The main task of the historian is to understand how individual people in the past have acted and why. The great unfinished business of social history, that is, of the kind of history that makes the full range of our dead predecessors its object of study, ought to be to give a *persistently* and systematically better account of the ordinary lives of the past. The lives themselves, not merely the *effects* of those lives: this means to figure out not only past agents' consequences, but also the *meanings* that guided people as they lived in and altered their worlds. We should interpret them as we might people we're speaking to, at least listening to, rather than merely explain them, as we might their diseases.

It is extremely difficult for even the best social historical intentions to catch and keep real people. However, if our concerns for understanding are to be met, we must focus on the self as readily as on the social. Given an imperative to capture both, how do we work with the following moments? In September 1385, Richard Ferrour, sometime Constable of Wells, sued John Benet, another citizen of Wells, for trespass. Then, as now, trespass was never a simple business matter. It disturbs the contours of the self, which you'll recognize if you've ever had your house or land invaded by strangers – a door ajar when you come home. At some level, it threatens the self. More banally, Ferrour specifically accused Benet of letting his animals eat Ferrour's standing grain. Ferrour insisted that it had happened 'many times'.[1] Typical of the Wells regime, the court appointed arbitrators, and they were a high-powered group in the little town. Ferrour chose the man who would become master of the town two days later, Nicholas Cristesham, while Benet selected the master of 1383, Henry Bowditch. Each disputant added a substantial but lesser man as his second arbitrator. This suit reveals the Ferrour–Benet relationship at a crossroads, a moment of explosive aggravation that would alter and intensify their public opposition and who they were in Wells. It also reveals the heart of social life, individuals glaring at each other, calculating their chances, suffering their emotions and their histories – all hard things for the historian to unearth.

Consider another moment, nestled in the social frame of the family. Across England, another burgess, John Kempe of Lynn, returned his wife Margery's buttery keys, symbols of her domestic and human powers, even though she had been recently violent, insane, terrible and possibly dangerous, and even though others in his household thought him foolish for doing so.[2] History must understand the world that enveloped such incidents without sacrificing the people to our analysis. Starting with the incidents near us we have to wonder which analytical tools will take us furthest in trying to understand what was going on between Ferrour and Benet or between the Kempes in front of their servants. While we may quickly suspect political ambition or economic rivalry operating in the Wells case and gender expectations in that of Bishop's Lynn, I want to suggest that to achieve the most requires us to make sure that the acting selves are not forgotten as we try to appreciate social processes.

Richard Ferrour is mentioned more often than any other person in later medieval Wells. He is someone the historian might be expected to understand: he was an active member of the elite of citizens of a town of regional importance. From 1369 until 1403, he shows up perhaps 300 times in the records, in a dizzying 200 legal entries. Yet we'll only understand him by thinking resolutely *beyond* his life, any particular incident and his own perspective towards the necessary conjunction of self and society. To pursue this most seriously – rather than as Marx's platitude of social history, 'Man makes his history but not out of whole cloth' – requires us to focus on selves and small groups and their particular interactions, even their ideas.[3] We need to focus on Richard Ferrour's *social self* to see what can be understood of a man whose numerous records don't even tell us where he was born or what his wife's name was. As for John Kempe, he is almost Ferrour's photographic negative in the documentary records, a man known more for his wife, Margery, than for himself. The Lynn burgess provides a second and complementary focus for understanding medieval people. To achieve something like understanding for difficult targets such as Ferrour and Kempe is to achieve it for a great many other people in their world, a clear social historical result. Through the social self, we can complete social history by imaginatively reconstructing the action world of historical actors.[4]

I use the concept of the social self to keep us on task.[5] The *social* self, the social *self*: not a dichotomy, not a binary opposition, not an afterthought, but the way people actually are in a room, in a group, in a conflict, in a world inevitably beside other people or thinking about them.

> And indeed there will be time
> To wonder, 'Do I dare?' and 'Do I dare?'

Time to turn back and descend the stair,
With a bald spot in the middle of my hair–
[They will say: 'How his hair is growing thin!'] . . .
My necktie rich and modest, but asserted by a simple pin–
[They will say: 'But how his arms and legs are thin']6

Probably less self-conscious than T.S. Eliot's Prufrock, Ferrour and Benet were under intense scrutiny by their arbitrators and witnesses. Their actions and words must have reflected it. The social stakes were high, as present and past masters watched them, just as they were for John Kempe when he had to decide whether to accept and certify that Margery had recovered and do so without losing too much standing with the servants who watched him and advised. We need to think constantly about how the social and the self, with the whiff of the private and unsocial, work together. For a social historian to do this requires focusing on how the individual helps to make up the society which simultaneously forms him or her. Culture is funnelled through the particular vision of participants, of people, of social selves, of Ferrour and Cristesham as they conferred behind the scenes, of John Kempe looking from wife to servant, thinking.

I don't mean to recommend that history be a collection of biographies or to propose a renewed methodological individualism. Thomas Carlyle held such a view – if we could mass the biographies of our heroes, all history would be done.7 However, even if we agreed with this theory, the task is impossible with heroes as numerous and world-historically irrelevant as small-town men, Ferrour and Kempe. In fact, in line with much recent thinking in social science, the history of the social self must stress the fuzzy relation of self and society.8 One of the most important reasons for this is because the social world begins within human inter-actions, not within individuals alone, nor from society. It is Ferrour and Benet; John and Margery Kempe; J. Alfred Prufrock ascending the stairs under the scrutiny of the ladies – each person so involved with the others that the lines between them, the distinctness of their ideas and bodies, are not always clear. Life is a series of trespass cases.

The idea in a nutshell is in Chaucer: 'So hadde I spoken with hem everi-chon/That I was of hir felaweshipe anon.'9 The *Canterbury Tales'* narrator becomes part of the company by *talking* with the others. And in so doing, he comes to appraise and characterize them, which is one of the social self's chief activities – the assessment of other people. Chaucer goes on to 'telle [us] al the condicioun of eche of hem' (l. 38). This means to categorize his fellows through already established social concepts, provided by the larger world through language and institutions. Crucially and fundamentally, everyone is to be reckoned using well-known categories of general use which are modified by the critical phrase 'so as it semed me' (l. 39). Note the ambiguity in this characteristic thought of the social self, *so as it seemed to me* – the self is unsure even as it is the mediator and manager of the broader culture's assumptions.[10]

We must follow Chaucer's lead to decide not only the social standing of participants but also the framework of social ideas in which people acted and imagined themselves. We must also pay the closest attention to the company people kept, their friends and enemies. Thus, we should start not by considering anything particular about Ferrour or Kempe or the cases at hand, but rather their various contexts, shared by many others – some of the background features that united all the players in each scene. At perhaps the highest level of generality, applicable as far as we can tell all across Europe, there is a gender context, a way of joining and separating men and women, womanly and manly, in thought and practice. It sometimes flares up like a volcano, as it had a couple of years earlier when Richard Ferrour had an altercation with a lower-status woman, Alice, and struck her, drawing blood.[11] Normally, however, gender is the earth's core, powerful magma, hidden, not easy to appraise. In certain respects, John Kempe may have had to struggle most with its implications, for ideas of husbandly dominance over *his* dependants were central to his struggle and social identity. When he threatened to rape his wife, he was urged by his social failure to rule his own roost, as well as by sexual frustration.[12] By contrast, Ferrour's entire world is evidence of the domain of gruff masculinity, the precise contours of which historians are just beginning to work out.[13] It is a man's world, so only certain possibilities are laid open for good and ill, and it is according to these that Kempe and Ferrour must act and judge. In the case of Ferrour versus Benet, there was no woman in sight, but gender is all around – the cooled basaltic flows of past volcanic activity, which had made arbitration a male preserve.[14]

Geographic context also matters to the social self, and this is England in the later fourteenth century; for Ferrour, it is the West Country and Somerset. It is urban, but small town; it is the city of Wells, a part of the country that John and Margery Kempe did not reach, but which in many respects reflected their own small town in provincial East Anglia, although their port town may have offered broader vistas, its trade links conspicuously involved with the Baltic and Germany.[15] As far as we can tell, Wells and Lynn, respectively, were their main social venues. Although Kempe sometimes followed his wife across the country and opted for a less community-bound existence than Ferrour, inevitably he came home to Lynn, and its attitudes and social exchanges limited what he could do.

To a greater or lesser extent, Benet, Ferrour and the other men acting in their dispute shared ideas, values and structures of belief beyond their common background assumptions of gender and urbanity. We cannot know what any of this meant to them unless we understand some of the cultural and cognitive baggage they carried. Indeed, the intellectual life of common folk has an importance for understanding the medieval world that social historians have never adequately recognized. For the most part their methods and theoretical assumptions have let them down so badly that the problem is sometimes invisible. British social history has tended to assume a pretty undifferentiated mental and cultural life for its subjects, aside from modest concerns for class consciousness. Historians and literary critics are now moving a long way beyond the helpful but limited work of historians such as Rodney Hilton or the even older work of Sylvia Thrupp, but much of the current work dealing with social ideas posits them not so much of people as of the culture as a whole.[16] I believe that the focus on the social self, a focus on understanding medieval people, will help a durable framework to emerge. Overall, we have made little consistent headway in thinking about how what was done was dominated by what could be thought. At the least, I want to make clear that understanding medieval people requires a more explicit consideration of what they thought or valued. As the six men faced each other in the arbitration hearings, weighing evidence and character, they brought ideas and assumptions to the scene which we need to specify before we can understand them. These ideas were not inert, wherever they came from. People made them act and changed them in the doing. The method of the social self requires the specification, as explicitly as possible, of the prevalent social notions, and then a sense for how individuals may vary the themes they all know.

The urban circumstance matters at this point. For while what was common to all England and all commoners is important, values mainly matter in action situations.[17] The place of action in a small town will bend the meaning of hierarchy or fidelity, friendship or equity, into a distinctive shape. Three values were so critical to the imaginary of urban social life, pungently crucial in Wells, that I suppose them to be mastering. These are honour, fidelity and hierarchy – together constituting the proper ordering of social relations. To simplify, these values respectively meant that Ferrour and Kempe would be worried about their relative standing in the community; that the faithfulness of their relations with others would constitute a major part of their social identity and the evaluation of social worthiness; and that everyone assumed a degree of stratification, notwithstanding the considerable fraternal impulses of an urban association such as Wells Borough Community. It matters less right now to figure out how such values are constituted, although cultural historians have shown how dominance and control were important considerations. For the social historian, it is usually enough to insist that the values ordain a certain kind of vision for social agents, denominate good and ill, but that they can't make choices and cannot (being a structure of ideas rather than people) *do* anything or *judge* anyone. This is what people are good at and what they are responsible for.

Without denying that my assertion of these urban master values may seem arbitrary here as I have no space to make a proof, I want to pursue them more concretely in the social world of Wells, alongside Richard Ferrour, by returning to his suit against Benet. The dynamic of this little case reflects many Wells traditions and assumptions. Arbitration itself is a window on honour. It is a dispute resolution technique that deals with people of notionally equal status.[18] King Richard might conceivably go to arbitration with King Philip, but he will not do so with Wat Tyler. In other words, it flourishes within a world of class equals. Arbitration is uncertain but trusting, by which I mean that the results are determined by others who are freely chosen by the disputants, who, in effect – and in fact by oath – commit their wills in advance to agreeing with the resolution. It is a free person's game and a matter, therefore, of honour. What is honour, which for a medieval woman seems so often to boil down to sexual respectability? Hobbes has the best definition: 'Honour consisteth in the inward thought, and opinion of the Power, and Goodness of another . . .'. The 'external signs' of that esteem were 'worship'.[19] To whatever degree people might be prideful or self-regarding as a matter of character, the social and cultural 'system' here re-enforces and cultivates these *particular* frames of reference. Citizens like Kempe and Ferrour *had* to understand themselves partly through the ways of honour. All civic officers, all citizens were to some degree worthy of worship.[20]

Fidelity joins honour as a central organizing structure of mind and practice, through which the medieval social self played. Fidelity indicates many things in the Middle Ages,[21] but it is ever the wish fulfilment of medieval social life, the hope/assumption of the support of others upon which people anchored all their plans. Today, we can lead with the emotions, for many of us have few outside a small knot of near family members to rely on, and, crucially, we *need* not rely on them. Contemporary social relations work as transient, efficient business transactions, whether with bureaucrats or in commerce. In the later Middle Ages, however, when it was time to ask men to arbitrate for your interests in front of the entire citizenship, you needed to rely on fidelity. This does not mean simple cronyism, although that *is* one way of fidelity. As often, it meant trust in a man's honour, in his need to act out and preserve *his* good name regardless of what he might think of you personally. Get the right man to agree and he'll do the job regardless of affection towards you. Get another 'right' man and he'll do anything *because* of his attachment to you. Either way, fidelity was an overarching master value, and men would do better if they learned to judge it and speak its language. When you were crossed or disappointed, an accusation of infidelity, of breaking your word, was a powerful social weapon, as well as an elevating self-righteous vent. The context of an idea is a large part of its meaning. In choosing Nicholas Cristesham, for instance, we can be sure that Ferrour was feeling all of this background pressure. He chose well, for Cristesham had supported him many times before and had already been borough master four times – a man prestigious enough to check Henry Bowditch, himself three times master.

For Ferrour and company, enough of their honour was at stake in the lawsuit to affect their social standing. Choosing their arbitrators reflected not only on the honour of all and imposed on the trustworthiness and fidelity of arbitrators, but it also mirrored the other master value of hierarchy or stratification. Wells and Lynn were virtually one-class towns, almost all bourgeois or bourgeois wannabes. This was categorically true among the citizens; consequently, however, people were almost obsessed by gradations of social standing. Thus, Margery Kempe was conscious of her father's superiority over her husband. Married though she was, she told the mayor of Leicester, 'I am of Lynne in Norfolk, a good mannys dowtyr of the same Lynne, which hath ben meyr five tymes of that worshipful burwgh.'[22] John Kempe was down the ladder and untactful as Margery was in pointing it out, she was right. I call this hierarchy, but it is close to what anthropologist Louis Dumont has called stratification.[23] For those with sharp eyes for the social self in Wells, all the men involved, disputants and arbitrators, could be arranged in a list from more to less worthy. Their honour was not merely a yes or no question for themselves, it was necessarily converted socially through this value of hierarchy into a current league table ranking: 1) Cristesham, 2) Bowditch, 3) Ferrour, 4) Benet, and so on. There was no class division, so nothing was settled and certain. Where exactly each individual fitted in needed to be taken into account at every point by all the participants. Moreover, how each person acted and spoke, including their comportment during a case of arbitration, affected their standing. At this point, however, the social listener would apply a whole range of other, lesser values and virtues to his assessment of his neighbour. It is a familiar process perhaps, but the very nature of an arbitration case intensified it: moral space was manufactured for Ferrour and friends. Indeed, the social self is always fundamentally bounded by its moral space.

In reflecting on this case, I have been examining the values that constituted the social atmosphere, the stage for social actors. Among burgesses and citizens, a locally inflected version of honour, fidelity and stratification (hierarchy) prevailed. We have also seen how concrete social relations mattered. These men were all together in a room, assessing each other and acting in company. We can do more with the Ferrour–Benet case because it highlights another feature of the social self that may seem most antithetical to an individualist approach. The social self should make us focus on the small-scale actions, perceptions and claims of others. To twist Hume, from the vantage of the social self, we are bundles of perceptions in other people's minds. Moreover, we are known and come to know ourselves by the company we keep. Again, this is highly sensitive to the particular historical and cultural moment, but in the medieval town (and the Middle Ages generally, I am sure) identity was greatly a public matter made by networks of friends and gazes of enmity. As John Kempe's reputation prospered by marrying the daughter, even the nutty daughter of Mayor Brunham, it fell with her reputation for disobedience and difficulty.

Let me broaden this important point, because it directs us to the power of friendship and enmity in medieval towns. The Ferrour–Benet case of 1385 was, in fact, the fourth court case that had brought them together. In 1379, they met as arbitrators in a case, but they worked for opposing sides.[24] February 1380 saw Ferrour's opponent, William Davy, choose Benet as his second arbitrator.[25] Late that year, in a case involving two other men, Ferrour and Benet were chosen to lead the opposing arbitration teams once again.[26] There is yet another instance: John Benet's first appearance in a court case was in December 1378 and his opponent was Henry Ferrour, Richard's brother.[27] Strikingly, their attitude towards each other was so well known that those whom Ferrour opposed knew John Benet was a worthy advocate. In other words, part of the social identity of each man was constructed from his enmity towards the other.

Of even more interest, however, are the practical consequences for third parties in this small-scale social world. Ferrour and Benet's antipathy was *counted* on by others in order to hedge their bets in their own lawsuits. This continued for several years after the 1385 case. Thus, one social self stretches out to mingle with others. At harvest time the following year, a servant of Robert Horn's trampled and destroyed some of Ferrour's grain, and when the case was sent to arbitration Horn chose Benet.[28] When in early 1387 Benet returned to court to sue Thomas Duk for back rent, Duk was quick to select Ferrour as one of his arbitrators.[29] And, just to show how such perceptions did linger in the public consciousness, there was the case of Thomas Smart versus James Masson in October 1388. In that case, Smart charged Masson with 'stealthily creeping onto his land' and trampling it, causing a variety of damage. The landowner chose Ferrour for an arbitrator; the accused trampler selected Benet.[30] Benet's reputation had been made. He was an important node in a social network that defined him and others.

Small, evanescent, but clearly operating social groups emerged from such well-known personal antipathies. Friendship was often a contextual relationship: shared antipathy or shared circumstance led men into each other's camps. People knew John Benet was honourable, but they also knew that he understood the difficulties of abutting a rich and litigious member of the oligarchy. He would be sympathetic. Plainly, the focus on an individual such as Ferrour ought never to be too narrow. To understand him is to understand others in his world, and it turns out this means to understand how lingering personal feelings acted out in public forums contributed to social life.

The study of social networks helps us to understand politics large and small. Ferrour's career provides general evidence for the way a man of some political importance in the town developed and sustained his power. I can't put a map of Richard Ferrour's social network on the page because it would be a mass as visually meaningless as the Milky Way, encompassing so much but yielding no manageable graphic pattern – except one. He was the man in everyone else's business and this both reflected and sustained his social self. Ferrour's vast social network was the basis for much of his influence. By contrast, a humbler burgess's social self was small, consisting of a few public performances and only occasional contact with powerful men. Such a man was likely to rely on craft and family connections. He turned inwards to his buddies, effectively circling the wagons rather than reaching outwards. Ferrour, however, was the supreme social extrovert and social power was friendly to such extroversion; his pattern was just an exaggeration of the norm for oligarchs.[31] Such men converted their honour and trustworthiness into political power by acting as the useful intermediaries between lesser men. Therefore, by arbitrating as Ferrour so often did (139 times!), he was entering lives and conversations, collecting information, as well as re-enforcing his role as universal friend, active in literally every other burgess's social network.[32] He was akin to a professional mourner. With the role came the emotions of friendship. And, unlike John Kempe's wife, Richard Ferrour was most often *invited* into people's business, and his success at working for the interests of others advanced his own standing. His social self literally spread across the community and helped to make it one.

Richard Ferrour's extroversion reflected not only his relative wealth and high estimation in the town, but his character and energy. Therefore, to understand Ferrour or Kempe we need to move closer to the individual and recognize more than we have so far the struggle of a self against the world. In other words, while lining up the social history, we also need to engage the personal, the contingent and the psychological. While there are a wide variety of theoretical means and psychological theories – from psychoanalytic to evolutionary psychological – that offer helpful frameworks, there is no reason to wed the theory of the social self particularly to any, even though there is a need to acknowledge the centrality of contingency, which psychological theories, even 'common sense' ones, generally suppose. I mean that we are required by the nature of social action and the social self to face the question of the difference that each individual (or chance) makes to how things work out. This is contingency. Sometimes we need to follow the record to see the character of a self to explain his traces. Some historians will want to engage a theory of character or personality to explain what they see. The 'appropriate theory will likely depend on the historian's inclinations and the kind of evidence at hand. The question that must arise is: What type of people are we dealing with?

Ferrour was the shrewd aggressor as well as an imperial, colonizing self. He made enemies, sued often, but never, save once, sued another member of the town's elite. In other words, he knew how and when to be aggressive. Even as an arbitrator, a man who had to work with others to make peace, he showed an aggressive character. He was in fact more likely, by a ratio of 3:2, to be the plaintiff's arbitrator than the defendant's. People knew he would press the matter home. His unique willingness to pursue others in the courts bears this out. It is less surprising in the end to find that he was the only oligarch on record as striking a woman. No other member of that elite went to court as often as he. His profile of sometimes angry activity limited his overall standing in the town. It was the blemish on his social self. Richard Ferrour never attained the master's chair and his *relative* lack of moderation may well have been the cause. There was a little too much edge, a little too much pride in his transactions and perceptions.

This was possibly because (more than any other Wells burgess of this period) Ferrour's social self was entangled with the physical manifestations of life: property and body, chattels and fields. As many scholars are showing, paying a lot more attention to self as body and self as possessor of property provides an important additional perspective.[33] To find traces of the self as agent of its own development and defence, look to the things you can touch or trample. Looking back on famous works of pre-modern individuality, such as *Montaillou* and *The Cheese and the Worms*, one can only be struck now by the fact that having so many ideas from the past suddenly bouncing around made us scarcely aware that people so often, even characteristically, think and feel through their things, including that most precious thing, the body.[34] Truly, those classics seem tales from an inquisition without torture.

In considering people, we need to know *where* their social selves were – a physical, even geographical perspective. For Richard Ferrour and John Kempe (and many later medieval people, at least), I would claim that they were in their bodies, in the persons of their family and relations, and to a greater or lesser extent, in their property, houses or fields, in their handiworks and other deeds. Even though they were very much present in their reputation or good name, the latter was really just a metonym for their social self's presence in things and places and people. The social self was, in other words, very much an infusion of spirit into matter. You put yourself into it, as we say.

Richard Ferrour's servants and fields represented him, while his vigorous, indeed excessive, action at law reflects the way that the culturally provided, socially constituted and economically certified importance of land and servant produced in him a kind of exaggerated effect. He seemed almost to fetishize these things. We have already seen, for instance, that Ferrour was at the centre of an almost factional dispute over trespassing, especially encroaching on fields. There was a violence in his nature, in his defence of a self spread across the townscape, which his position generally let him express in a licit, even laudable manner as an honest, honourable busybody. But when he struck Alice that night he let us see something else. Striking subordinates was often acceptable, but not to the point of making them bleed. He doubtless had better moments, but he was a world of manner away from John Kempe, who could not rule his wife and whose debts depressed him so much he needed his wife to clear them. From one perspective, Kempe was almost laughably weak, as when he trades his debts, including the insistent marital debt of sex, for Margery's money and her company at Friday dinner.[35]

Kempe's actions traded his honour away. He showed himself reliant on his wife, and his social standing withered as she publicly discharged his debts, a public version and objective correlative of his sexual humbling. Contrasting Kempe's decline, his relinquishing control of Margery's body, there is Ferrour's constant advance in the public's face, if not always their esteem. Ferrour sued to protect his crops and lands with a proud persistence, but he was also an encroacher on others. This included, in 1382, storing his goods illegally in the public streets, taking a bit of the town over as if it belonged to him alone.[36] Similarly, in 1390 he was cited for allowing or putting some of his cattle on the local lord's land at nearby Burcott.[37] Bearing in mind his earlier disputes, we should see Ferrour as a most physically formed social self, an avid owner, a marker of fence posts, probably a *provocateur*. When others fought back, he fought harder, so sometimes feelings hardened and disputes lingered. In 1396, he accused his neighbour, John Pestel, of obstructing his gutter with the flotsam of woad dye.[38] To Laurence Iforde he complained of a pig trespassing on his property.[39] In 1382, Ralph Tucker could-n't keep gates closed conscientiously and Ferrour's barley suffered as a result.[40] Eight years later, William Wynd also couldn't keep the close sealed off, to Ferrour's damage and continuing irritation. In the way of the expansive social self, Ferrour's agency was also in his servants, who learned their manners from their master. One servant so enflamed a neighbour, Agnes Knyght, that she attacked him for his impertinence. Ferrour sued, for as all family members, servants were part of a person's social self, a place he could be hurt or honoured, part of him and under his control.[41] While many of these suits were of course of economic significance – what isn't? – the concern for Ferrour and his world was the defence of his integrity. Mappable and palpable, his soul clung to barley and byre, walls and woods and wool. And he was not good at sharing.

The same melody is discernible even in John Kempe, although it is played in a very distant key. The greatest accoutrement a man had was his wife and to possess Margery Kempe was no small thing. That Margery resisted this possession does not remove the fact that they were tightly bound together. When a mob called to burn Margery Kempe, 'and the creatur stod stylle, tremelyng and whakyng ful sor in hir flesch wythowtyn ony erdly comfort, and *wyst not wher hyr husbond was become*', John Kempe was failing and shrinking in public.[42] She was a part of his social self, and in fear and embarrassment, he sometimes disappointed her and ran off, fearful of losing too much. This isn't to deny (and she doesn't), that he was her most steadfast supporter, but he was not a hero for all that.

There may have been hopes that he would be such an urban hero. His father had been, and while it is possible that reverses in the Baltic hurt the family's prospects even before John Kempe's marriage, his path in Lynn was neither smooth nor clearly upwards.[43] Strikingly, he was sworn into public office in 1395 as one of the greater council – what one would expect of John Kempe Sr's son and John Brunham's son-in-law – but, embarrassingly, he had to resign the office a few months later and he never again held any other office.[44] Something had miscarried with very public consequences. His social self never really recovered, it would appear. Economic decline and frustration marked his identity.

Another way to appreciate his business reverses and strong-willed wife is to see that for much of the time, a large part of John Kempe's social self was out of control. Social selves, so involved with other people, are inevitably split personalities, but the problem was acute with John Kempe, who was beleaguered by a resistant wife, servants and debt, which is also a kind of reduction or alienation of self. The cost of Margery's early fine clothes may have bothered John, but both must have known that clothes were one of the main means of expressing the social self. John Kempe worried over his wife's 'pompows aray', whose pomposity was its inappropriateness, a straining for great social effect. The strain showed, for the social self embedded in a hierarchical world can only get the honour fitting to it, and to her pain, Margery Kempe was stuck with John Kempe's social self, which limited possibility. Later she asked his forgiveness for ignoring his advice to dress more modestly.[45] Even later, he would drop all such questioning of her clothing, when he came to read her through a register different from that of the Richard Ferrours of the world. But at every point of her self-expression, Margery was also expressing John Kempe, whose control over himself was thus unusually compromised. He was there when Marjorie had a go at showy clothing, at brewing and milling, at fornication and pilgrimage, and at revelation – there, even when he was physically elsewhere. She spread the humble Kempe across large social swaths of the country in puzzling and troubling guises. Not all venues of expression will be as prestigious to all eyes. Richard Ferrour's ways with real property and agrarian pursuits were less suspect, I suppose, to townsfolk than Marjorie Kempe's holy afflictions would have been. Mrs Ferrour left no marks, whereas Margery's religious path was a marginal pursuit, a long-shot route to social prestige, but a surefire path to a large, if lonely, social self, trailed by her husband John. Why people choose one venue of expression rather than another is partly a matter of circumstance and the luck of life – gender and inheritance, physical stature and cleverness – partly a matter of convenience and partly one of personal preference, including the outing of our psychological natures.

This is all gathered together in a reputation. At every point, the pursuit of a certain kind of social identity and even of wealth beyond mere subsistence is an attempt to secure a certain image, a certain social self. This means that you defend that image against slanders, against the opinions of others. You buy back some of your own company's stock. You launch defamation suits. The defence of a good name was a constant activity in medieval and early modern societies.[46] Wells evidence is full of such cases. Unsurprisingly, this is also the era of the Scottish poetic genre of flyttyng, in which men have a competition of putting each other down.[47] I suspect Richard Ferrour was put down only in select company of enemies (perhaps at John Benet's table), for Ferrour's punishing litigiousness – and the suspicion that he was perhaps charming as well as vindictive – protected him from actual charges. His good name stood.

John Kempe, by contrast, spent a lot of time suffering a bad reputation. His wife was a part of him that humiliated him in the early years, through instability and unhappiness with their social standing. He hurt himself by his indebtedness and his public obligation to his wife, who even after all her businesses went bust still had the capital to buy out *his* creditors. His trajectory in Lynn was not conventionally upwards. What was it, however? For John Kempe shows a remarkable transition, which alerts us to the fact that, alongside the values of Richard Ferrour, there existed a host of interweaving and contradictory hopes, some potent inversions of the standard burgess's goals. Against the imperial self of Ferrour was the recessive self of John Kempe, a man who went out on pilgrimage with his wife, increasingly poor, increasingly inward, increasingly religious. For, whatever else we may see in Kempe, we should see how susceptible he was to his wife. She came to master him in most respects. The later humiliations and pains were related to the earlier. They came in suffering the abuse Margery received from those who retained secular influence or out of simple human viciousness. When in Lambeth an apparently worthy matron heckled Margery and looked forward to feeding her heretic's fire at Smithfield, John Kempe 'suffred wyth gret peyn and was ful sory to heryn hys wyfe so rebukyd'.[48]

Margery's charismatic mastery of him, however, produced John Kempe's conversion from the secular values of Richard Ferrour to the religious values of humiliation – social and otherwise. It was not without social purchase either. After he started following his wife, Kempe's life acquired the cast of religion, which was the cast of power too. Such power must have made sex positively frightening for him, encouraging their eventual vows of chastity. Moreover, at that time he broke out of the local world of Lynn and was refreshed, possibly even revolutionized by the company of clerics and mystics, and by the reflected attention of bishops, doctors and persecutors. Amidst the admiring clerics, Kempe 'had ther rygth gret cher . . . becawse of hir'.[49] He was there with Margery, when she, like Yeats' Crazy Jane, talked to the bishops, reading them her lectures. Even at such dangerous moments, pride in his wife must have offset the fear and humiliation of other times, *his* failures, when he 'went away fro hir as he had not a knowyn hir', out of 'veyn dred'.[50] However, we may sell him short even here, for once committed to the cause of the chaste life, he seems not to have flagged. The text says '*he* led *hir* [my emphasis] to spekyn with the Bysshop of Lynkoln', Philip de Repindon, to try to formalize *their* profession of chastity.

He was not only her follower but her companion, and perhaps her apostle too. For if she struggled to learn from God that everything was working out, John Kempe had come to believe in their mission and in this general approach, notwithstanding adventures and threats that Voltaire's Candide alone could have embraced hopefully. John Kempe was 'alwey trostyng that al was for the best and schuld comyn to good ende whan God wold'.[51] Such a stoical approach to life can be debilitating when faced with the fire, anger and sarcasm of Richard Ferrour and the urban elites, perhaps even of his father-in-law, John Brunham; but once such a man turns towards the real distinguished thing of religion and persecution on the road – becomes, in other words, a bit like St Paul – he finds himself tolerably well armed to cope. John and Margery Kempe had withdrawn from Richard Ferrour's game and John may have come into his own in the broader community of humiliating Christianity. By the time that Margery actually set off for the Holy Land, Kempe may have cared less that part of the procedure was the announcement from the pulpit that all his debtors, as well as hers, should present their bills to *her* before she left.[52]

The great irony for John Kempe and many a social self was that the end of the affair leaves them so close to a body and its accoutrements, a world including wife and family and house, but without an active self. Kempe got old, fell down a flight of stairs, and found his body broken and never right again. Not long after, his mind withered, turned childish and lacked reason.[53] Margery came back home and took care of him, with all the ambivalence of long-term caretakers. He received the labour of her body once more. But what was left of his social self was a settled reputation, adjusted only by the success of his family. At this point, we could agree with those scholars who see agency as less relevant than social structure, for John Kempe, though alive, though a social self, had ceased to be a social actor.

I can make a character sketch of Richard Ferrour too, approximating biography, beginning something like this: He saw almost everything of significance that happened for forty years, his eyes on everyone, an ear as filled as the priest's with other people's problems and the awkward details of their past. He could be a kind of emblem of the medieval townsman. For Richard Ferrour was a success. He was a long-standing member of the judicial leadership, but could not quite keep the eager and anxious self from upsetting people and from upsetting himself. He could get angry and even violent. Age did not mellow him even if wealth insulated him from its most grotesque social forms. Yet, he was the greatest of friends and had the whole world as his familiar acquaintance, probably bringing some to the point of love and certainly leading others, for a time at least, beyond the boundary of hatred.

What I'm trying to argue, however, is that such a sketch and the ability to make Ferrour or Kempe individuals is only part of developing a version of sociocultural history that puts people and the meanings of their lives at the centre. The social self is really just my term for providing a focus for doing this, which starts with realizing that the complexity of people means they both constitute their society and simultaneously change it. Historians, however, have rarely worked this out in detail. They have often believed, especially in social history, that people could be counted on to stay analytically simple. In some versions this was because life was thought to be so hard, so pressed up against scrabbling for a living that people didn't have time for much more.[54] This bestializing of people – and most mammals do make more of life than some people's versions of medieval peasants – makes little sense. There is always time behind the plough or in the factory for a little fantasy or a little philosophy. When Margery Kempe's miller ran off from her employ, it was because he knew there was meaning in the horse's refusal to work and a worrisome significance for him in his remaining with Margery. When John Kempe decided to stick with her, more or less, it was because he was convinced, through terror and hope, that it was ordained and important.[55]

What did it mean for them? This is the crucial question. To answer this is not to abandon discussions of historical change, but to try to link up with other discussions which have been struggling to ground speculations about such change in the way Europeans have altered as selves, as egos and as social actors. What did it mean for them? To some extent, historians of ideas, intellectual historians and students of *mentalité* and culture have answered this question.[56] My approach, however, argues that facing the question of meaning is not just about 'ideas'. The right answer cannot be simply an intellectual one, nor purely psychological. The right answer will be a compound of approaches that develops from considering that social meaning exists only in social individuals, in the social action between people. No one creates language alone, nor any other pattern of culture. It's a co-op. Ideas come down from Mum and Thomas Aquinas; they come in contexts; they come with commentary, praise or ridicule. They are twisted, sometimes for good and sometimes for ill. People remake meaning, sometimes intentionally, usually on the fly, but always based on the practices and meanings they already know. The place to start is with the social person as a crossroads of practices and ideas, a little market square of culture, firmly in the dusty material world.

The person, however, is not disinterested. Prejudices and pride are fundamental and the self is a shorthand for these and our odd, unstable form of self-awareness and interest. It is my approach's prejudice to see this social self, the processor of meanings and the only possible actor, as the analytic centre. At the same time, however, I recognize the fundamental susceptibility of people to others. Hence the importance of social processes, infused with forms of meaning such as hierarchy, reputation and honour; it is as crucial, however, that we discover the importance of *actual* linkages to family and friends, networks and factions. Putting all of this together decidedly tells us something about the social world under examination, but it ends with Richard Ferrour and John Kempe rather than with an abstraction towards society. We might abandon society, as Michael Mann has suggested, without abandoning the social.[57] Indeed, if I'm right, the more we look at the self, the more we will learn about the social. By resolutely *interpreting* past agents, we may come to more satisfying explanations of past times and learn how people in very similar worlds, aware of the same values and powers – people like Kempe and Ferrour – could turn in very opposite directions and so come out in such different places.

Guide to further reading

Clarissa Atkinson, *Mystic and Pilgrim: The Book and the World of Margery Kempe* (Ithaca, NY, 1983).

Anthony Giddens, *The Constitution of Society* (Cambridge, 1984).

Anthony Goodman, 'The Piety of John Brunham's Daughter of Lynn', in Derek Baker (ed.), *Medieval Women: Essays Dedicated and Presented to Professor Rosalind M.T. Hill* (Oxford, 1978), pp. 347–58.

Laura Gowing, *Domestic Dangers: Women, Words and Sex in Early Modern London* (Oxford, 1996).

Richard Firth Green, *A Crisis of Truth* (Philadelphia, PA, 1999).

Barbara Hanawalt, *Of Good and Ill Repute. Gender and Social Control* (New York and Oxford, 1998).

R.H. Hilton, *English and French Towns in Feudal Society* (Cambridge, 1992).

Rosemary Horrox (ed.), *Fifteenth-Century Attitudes* (Cambridge, 1994).

Karma Lochrie, *Margery Kempe and Translations of the Flesh* (Philadelphia, PA, 1991).

Sanford Brown Meech (ed.), *Book of Margery Kempe*, Early English Text Society O.S. 212 (London, 1940).

Derek Neal, 'Suits Make the Man: Masculinity in Two English Law Courts, c.1500', *Canadian Journal of History* 37 (2002), pp. 1–22.

D.M. Owen, *The Making of King's Lynn* (Oxford, 1984).

Nancy F. Partner, 'Reading the Book of Margery Kempe', *Exemplaria* 3 (1991), pp. 29–66.

Julian Pitt-Rivers, *The Fate of Shechem, or the Politics of Sex* (Cambridge, 1977).

L.R. Poos, 'Sex, Lies and the Church Courts in Pre-Reformation England', *Journal of Interdisciplinary History* 25 (1995), pp. 585–607.

Edward Powell, 'Settlement of Disputes by Arbitration in Fifteenth-Century England', *Law and History Review* 2 (1984), pp. 21–43.

David Gary Shaw, *The Creation of a Community: The City of Wells in the Middle Ages* (Oxford, 1993).

David Gary Shaw, *Necessary Conjunctions. The Social Self in Medieval England* (New York and London, forthcoming).

Charity Scott Stokes, 'Margery Kempe: Her Life and the Early History of Her Book', *Mystics Quarterly* 25 (1999), pp. 10–66.

Sylvia L. Thrupp, *The Merchant Class of Medieval London* (Ann Arbor, MI, 1948).

Transactions of the Royal Historical Society (6th Series) VI (1996), pp. 137–245.

Karl Weintraub, *The Value of the Individual* (Chicago, IL,1978).

Notes

1   Wells Town Hall, Convocation Book I (hereafter CBI), 58.
2   Sanford Brown Meech (ed.), *Book of Margery Kempe* (hereafter *BMK*), Early English Text Society O.S. 212 (London, 1940), pp. 8–9.
3   'Eighteenth Brumaire of Louis Napoleon', in John B. Halsted (ed.), *Contemporary Writings on the Coup d'Etat of Louis Napoleon* (New York, 1972), pp. 141–2.
4   William Reddy, 'The Logic of Action: Indeterminacy, Emotion and Historical Narrative', *History and Theory* 40 (2001), pp. 10–33.
5   The term seems to have started with George H. Mead, *Mind, Self and Society*, ed. Charles Morris (Chicago, IL,1934), pp. 178–86, 200–9, and *The Individual and the Social Self*, ed. David L. Miller (Chicago, IL,1982).
6   From T.S. Eliot, 'The Love Song of J. Alfred Prufrock', *The Complete Poems and Plays* (New York, 1971), p. 4.
7   *On Heroes, Hero-Worship, and the Heroic in History* (London, 1897), p. 1.
8   Cf. Anthony Giddens, *The Constitution of Society* (Cambridge, 1984), and Brian Fay, *Contemporary Philosophy of Social Science* (Cambridge, MA, 1996).
9   'General Prologue', *The Canterbury Tales*, lines 31–2.
10 Cf. Elizabeth Deeds Ermarth, 'Agency in the Discursive Condition', *History and Theory* 40 (2001), pp. 34–58.

11 Lambeth Palace Library (London), Estate Document 1182 (hereafter LPL ED).
12 *BMK*, p. 23.
13 Cf. Derek Neal, below (ch. 9).
14 It plays a considerable role, however, in the larger project: see David Gary Shaw, *Necessary Conjunctions. The Social Self in Medieval England* (New York and London, forthcoming).
15 John Kempe's father was involved in this trade: see D.M. Owen, *The Making of King's Lynn* (Oxford, 1984), pp. 332–3; and Margery travelled there, following her son, *BMK*, p. 229.
16 See, for example, R.H. Hilton, *English and French Towns in Feudal Society* (Cambridge, 1992), and Sylvia L. Thrupp, *The Merchant Class of Medieval London* (Ann Arbor, MI, 1948); but see also Rosemary Horrox (ed.), *Fifteenth-Century Attitudes* (Cambridge, 1994), and Barbara Hanawalt, *Growing Up in Medieval London* (New York, 1993), for balanced approaches.
17 For a full analysis, see Reddy, 'The Logic of Action', pp. 10–33.
18 See Edward Powell, 'Settlement of Disputes by Arbitration in Fifteenth-Century England', *Law and History Review* 2 (1984), pp. 21–43. On my underlying thinking, see the classic Julian Pitt-Rivers, *The Fate of Shechem, or the Politics of Sex* (Cambridge, 1977), pp. 1–47; *Transactions of the Royal Historical Society* (6th Series), VI (1996), pp. 137–245.
19 Thomas Hobbes, *Leviathan*, ed. Richard Tuck (Cambridge, 1991, orig. 1651), p. 248.
20 Needless to say, this is not proved by this case but illustrated by it.
21 See Richard Firth Green's survey of the related term 'truth' in *A Crisis of Truth* (Philadelphia, PA, 1999).
22 *BMK*, p. 111.
23 See his *Homo Hierarchicus*, trans. Mark Sainsbury (Chicago, IL, 1970), esp. Appendix, pp. 239–58.
24 CBI, 44.
25 CBI, 49.
26 CBI, 29.
27 CBI, 40.
28 CBI, 65.
29 CBI, 68.
30 CBI, 73.
31 I have developed this further in *Necessary Conjunctions. The Social Self in Later Medieval England* (New York and London, forthcoming).
32 This includes indirect contacts, a friend of a friend, for instance.
33 See Natasha Korda, *Shakespeare's Domestic Economies* (Philadelphia, PA, 2002), for example.

34 Emmanuel Le Roy Ladurie, *Montaillou*, trans. Barbara Bray (London, 1978), and Carlo Ginzburg, *The Cheese and the Worms*, trans. John and Anne Tedeschi (Harmondsworth, 1980).
35 *BMK*, pp. 23–5.
36 LPL ED1181.
37 LPL ED 1184/6; this was the Bishop of Bath and Wells.
38 CBI, 121.
39 CBI, 33.
40 CBI, 12.
41 CBI, 42.
42 *BMK*, p. 28 (my emphasis).
43 D.M. Owen, *The Making of King's Lynn* (Oxford, 1984), p. 332, which notes his father's difficulties over merchandise seized in Germany in the mid-1380s.
44 From King's Lynn records (KL C10/1 (RR), f. 120 and 124), as noted in Charity Scott Stokes, 'Margery Kempe: Her Life and the Early History of Her Book', *Mystics Quarterly* 25 (1999), p. 65.
45 *BMK*, pp. 9–10.
46 See, for instance, L.R. Poos, 'Sex, Lies and the Church Courts in Pre-Reformation England', *Journal of Interdisciplinary History* 25 (1995), pp. 585–607, and Laura Gowing, *Domestic Dangers: Women, Words and Sex in Early Modern London* (Oxford, 1996).
47 See William Dunbar, 'The Flyting of Dunbar and Kennedie', *The Poems of William Dunbar*, vol. I, ed. Priscilla Bawcutt (Glasgow, 1998), pp. 200–1.
48 *BMK*, p. 36.
49 *BMK*, p. 37.
50 *BMK*, pp. 27 and 32.
51 *BMK*, p. 33.
52 *BMK*, p. 60.
53 This is paraphrased from *BMK*, p. 181.
54 Natalie Z. Davis, 'Boundaries and the Sense of Self in Sixteenth-Century France', in Thomas C. Heller, Morton Sosna and David E. Wellerby (eds), *Reconstructing Individualism. Autonomy, Individuality and the Self in Western Thought* (Stanford, CA,1986), p. 53.
55 *BMK*, p. 10.
56 Cf. Karl Weintraub, *The Value of the Individual* (Chicago, IL, 1978).
57 Michael Mann, *The Sources of Social Power* (Cambridge, 1986).

2

Biography and autobiography in the Middle Ages

Jay Rubenstein

Biography and autobiography are modern categories. Most of the critical attention which they have attracted – autobiography especially – has concerned itself with the problem of how and why modernity gave birth to them at the Middle Ages' end.[1] Historiographical responses to the particular problem of medieval biography and autobiography have produced, for the most part, a few well-worn commonplaces. Medieval writers did not produce proper biographies, in the sense that they were setting out to explain what made their subject unique or unusual or particularly compelling. Instead, pre-modern writers produced their *vitae* according to set models or patterns. Ruler panegyrics, like Asser's *Life of Alfred,* are imitations of Einhard's *Life of Charlemagne,* itself an imitation of Suetonius. Writers of saints' Lives tried to show how their saints were like other saints, especially St Martin of Tours, whose biography by Sulpicius Severus established Europe's hagiographic agenda for the next millennium, and St Martin simply wanted to be like Christ. Biographical writings do tend to become more personal and more ambitious in the twelfth century. At the same time, remarkable examples of autobiography suddenly appear. But medievalists have been less interested in these textual changes than with grander issues, such as whether the twelfth century witnessed the 'discovery of the individual'.[2] Textual sophistication, apparently, is beyond the grasp of medieval writers themselves; they can only act as pawns of forces which are beyond their grasp and which we can define with only maddening inexactitude.

One of the most ambitious of these twelfth-century 'intimate biographies', to use the expression of Sir Richard Southern, is *The Life of St Hugh of Lincoln* by Adam of Eynsham. Like many medieval biographers and chroniclers, Adam seasons his text with occasional autobiographical interludes, including a description of a remarkable dream, which he experienced just before Hugh's death, though in his text he describes the vision immediately after the death itself. In the vision, Adam stood in a garden outside the bishop's house. To the south of the garden was a ditch, and beyond the ditch, a cemetery. Stretched out across the garden and extending into the cemetery was an enormous pear tree. 'Whoever has seen such a beautiful tree?' Adam wondered to himself. What a pity that it should stand there in decay, untended, when its wood could have provided writing tablets to all the scholars in England and France. And then, unexpectedly, Adam placed his own arms around the tree and lifted it from the ground. As he did, all branches and leaves fell from it, leaving in Adam's arms only the trunk. Adam awoke the next morning and revealed the dream to those he deemed worthy. Its immediate import was obvious: Bishop Hugh was going to die. The real significance of the dream, however, such as why Adam was able to lift the tree so easily, did not become apparent until, in Adam's own words, he 'had written this homely but unpolished account of Hugh's life and virtues'. The branches and the leaves from the tree are the leaves of the book, the *Life of St Hugh*, those few and comparatively insignificant memories which the dreamer and author, Adam of Eynsham, has sent into the world. The tree trunk, perhaps surprisingly, is not Hugh himself, or the actual life of Hugh, from which Adam has shaken anecdotes and exempla. Rather, again in Adam's words, it is 'the writer's memory' of Hugh, of which only a few small fragments can be shared.[3]

The vision is as remarkable for its beauty as for its symbolic sophistication (Adam in fact gives it one other interpretation to which I shall return at the end of this essay). Adam recognizes in it that he cannot give his readers Hugh's life as the saint actually led it. He can give only his memory of Hugh, and an imperfect realization of his memory at that. It is arguably a more sophisticated understanding of the relationships among text, self and subject than the methods used by more recent historians. A biography does not open a window onto its subject. It is instead pieces of memory, whose connection to a historical reality is elusive. While not itself a theory of biography, Adam's dream does make the crucial point at the heart of this essay. The life depicted in a biography is ultimately inseparable from the life of the biographer. All writing may not be autobiography, as others have suggested, but all biography to some extent is autobiography.[4] These autobiographical elements can be personal and direct – Adam of Eynsham, for example, spent all but one day of the last three years of Hugh's life with the bishop and, as a result, is himself a significant character in Hugh's story[5] – or they can be more indefinite, the result of something akin to modern theoretical notions. The narrative of a person's life grows out of beliefs of how a life ought to be lived, and a person's beliefs about how a life ought to be lived inevitably relate to that person's own life. It is surely no coincidence that the revolution in biographical writing which occurred around the beginning of the twelfth century happened at precisely the same point that medieval people began to experiment with autobiography.

Stated most simply, people like Adam of Eynsham had begun to think in new ways about what it meant to live a life. Traditional approaches, which have emphasized the ascendancy of group identities in medieval culture, have downplayed the significance, or even the existence, of the self in the medieval imagination.[6] But as we shall see, medieval thinkers, especially from the twelfth century onwards, frequently did think in abstract and creative ways about the self, and biographical and autobiographical writings presented ideal fora for the practical application of these speculations.

Before we attempt to apply or impose any modern theoretical system of interpretation to medieval narratives, I should emphasize again a crucial point: medieval writers had to hand their own narrative and interpretive systems. Sometimes a particularly senstive or sympathetic reader can tease out the outlines of what those models were. Benedicta Ward, for example, presents a convincing case that the Venerable Bede fitted the five books of his *Ecclesiastical History of English People* into the model of the first five ages of the world.[7] The possibility that anyone will ever be able to discover such a framework in the *Historia* of Gregory of Tours, on the other hand, seems unlikely.[8] Gregory himself apparently saw connections otherwise invisible to modern readers, given the regularity with which he begins chapters, regardless of chronological or geographic connection, with the words, 'The next thing that happened . . .'. By contrast, Adam of Eynsham – whose book is remarkably coherent by comparison – often apologizes to his readers about the unusual form that his biography has taken. In his introduction he regrets that he has not written a full and sequential narrative of Hugh's life (*non seriatim et integre*), even though, to all appearances, that is exactly what he has done.[9] Perhaps no one captures better the confusion and the monumental creativity required to produce a fully realized biography than does Eadmer, the disciple and biographer of Saint Anselm of Canterbury and Bec. In order to capture something of what he believed about his master, Eadmer wrote not one but two books, the *Historia novorum*, in which he intended to describe Anselm's public, political life, and the *Life and Conversation of Anselm, Archbishop of Canterbury*, with which he intended to capture Anselm's private character. The form is unusual and unwieldy, and when Eadmer advises his readers on how to approach his texts, he gives precisely contradictory advice. Neither of the works, he says, 'stands much in need of the other for its understanding.' And then he immediately warns his audience that they cannot hope to understand Anselm unless they read both books.[10]

Any attempt to tease out medieval theories and models is made all the more frustrating by the apparent clutter of the medieval authorial mindset. It is an outlook which must have resulted in part from the kind of academic training which medieval writers experienced in that most central exercise in their curriculum, the study of the Bible.[11] Scripture, to the medieval mind, functioned on a variety of interpretive levels, most commonly four – the literal (what the words actually say), allegoric (how the words comment on the development of the Church), tropologic (what advice the words give on how to live a moral life) and anagogic (how the words foretell the Last Days). The dense and unattractive biblical glosses and exegeses reflect the true medieval reading experience, an experience from which no single, coherent narrative can emerge. One does find in the Bible a simple, literal story – that of the world's Creation and redemption – but it is not a narrative one simply reads through. One instead pauses over each word, draws from it as many meanings as possible, stopping occasionally to consider longer passages and how the various larger stories might fit together.[12] We can easily recognize the significance of biblical exegesis if we simply remember that the majority of work by the Middle Ages' first autobiographer, Guibert of Nogent, was exegetical. We can see its influence just as readily in the writing of Adam of Eynsham, who litters his text with an array of biblical analogies. About monks who abandon Hugh's monastery, he observes that, 'like Lot they departed from the steep mountain of contemplation, to be saved maybe in the Segor of good works. Unlike Moses they were unable to enter the cloud in which God was, but returned to their tents where they had fomerly dwelt.'[13] Sermon styles and storytelling – whether simple anecdote, the narrative of a life, or the history of a nation – inevitably will reflect these habits of scattered thought.

The *vitae* of early medieval saints follow this exegetical pattern on a very basic level.[14] They seek on a literal level to relate particular events from the life of a saint. On an allegoric level, they show how the pattern followed by their saint is the same as that followed by their saintly predecessors who, as noted above, imitated the life of Christ. The saint is born to the accompaniment of heavenly signs. After a brief and stylized childhood, in which the saint, like Christ, is always marked as distinct from other children, we hear about a conversion to the ecclesiastical life, triumphs over devils and demonic snares, miracles performed and, eventually, death, which, like the saint's birth, is accompanied by heavenly signs. It is a pattern with room for variety. The conversion can be Martin's renunciation of the Roman army, Benedict abandoning his studies for a hermitage or Guthlac leaving a Germanic warrior band for the monastic life.[15] The gradual promotion in virtues, generally revealed through miracles and portents, can play itself out through a career in court, as with St Dunstan of Canterbury, or with the suffering that accompanies a life in ecclesiastical politics, as with St Wilfrid of York.[16] Whatever the variations, these *vitae* share two important characteristics, which together tend to limit the sense of a 'self' within the text. First, they show a life lived at one pitch. Even though the story includes a moment of conversion, readers never need doubt the core virtue of the protagonist and need feel no uncertainty as to the eventual outcome of events. Second, the pattern of the *vitae*, whatever the variations, is essentially Augustinian. Life is a pilgrimage. The saint is constantly in a state of forward motion, walking from the City of Man to the City of God. Their characters are static and the paths they walk, though filled with obstacles, neither fork nor break.

Autobiography is necessarily a markedly different project. The model for medieval autobiography is obviously St Augustine's *Confessions,* a book whose narrative is most assuredly not Augustinian in the sense described above. It is, to borrow a phrase from Peter Brown, the story of 'the evolution of the "heart"', which is 'the real stuff of autobiography.'[17] Augustine's purpose is to describe the soul's motion towards God, but it is not a straightforward and continuous movement. It is, instead, a road full of dead ends and detours. Even by the end of the book, after Augustine's full conversion, he lets his readers know that his soul's ascent to God is by no means complete. Consider his description of his sexual urges: 'When I am awake they beset me though with no great power, but in sleep not only seeming pleasant, but even to the point of consent and the likeness of the act itself.'[18] It was a sentiment rediscovered by Guibert of Nogent, the eleventh-century intellectual heir of Augustine who would be the first western writer in more than seven centuries to attempt an autobiographical exercise. Concupiscence, Guibert would observe, survives tenaciously, even among the old. 'What evil is more bitter, more untamed! Oh, for shame!'[19] It is a far cry from sexuality as seen in a traditional *vita,* such as the 'Life of St. Benedict'. There we read that Benedict overcame carnal urges by hurling himself into a thorn bush and badly cutting up his flesh. Afterwards he never again felt such heat.[20] One extreme act of asceticism banished sex from his thoughts. An autobiographer like Augustine or his later medieval successors did not share in the luxury of being able to declare their battles won.

This observation points to the most basic and obvious difference between biography and autobiography, the difference which makes the concept of autobiography almost impossible with traditional medieval concepts of how to tell the story of life. Simply put, an autobiographer is not dead. The story of a life makes no theoretical sense if it is not leading towards a death and the completion of the soul's pilgrimage to God. Augustine overcame this difficulty by making a sudden inward turn in Book X of his *Confessions,* but the sophistication behind his speculations was beyond the skill of most writers. Before the twelfth century one finds, instead of developed autobiographies, 'autobiographical moments'. Einhard, in the introduction to his life of Charlemagne, takes care to note his friendship with Charlemagne, Charlemagne's charity towards him and his ongoing connections with Charlemagne's children. Similarly, Osbern of Canterbury, in his *Vita S. Dunstani,* describes how he himself had once visited Dunstan's cell at Glastonbury. 'Now that I have viewed the holy place where he lived, I confess myself a miserable sinner! I have seen the works of his hands and touched them with my own sinner's hands. I turned my eyes to them, cried out rivers of tears, and worshipped them on bended knee.'[21] Autobiographical moments can also be subtle. Both Osbern and Eadmer, when writing about St Dunstan, apparently include miracle stories about themselves, though they do not identify themselves explicitly as actors.[22] Osbern and Eadmer as well each had personal agenda behind the production of their hagiographies – namely how to stamp the hagiographic and liturgical practices of their community with their own particular vision, in opposition to their former archbishop Lanfranc, who apparently sought to downplay the importance of his community's relic collection and whose reforms had long-standing repurcussions.[23] Moments such as these are an imperfect solution for what must have been for medieval historians and biographers a frequent urge and an impossible dilemma – how to make the story of others' lives into the story of the writer's own life. The autobiographical urge, the wish to place one's self within the text, is a desire and habit which long pre-dates Guibert of Nogent's rediscovery of the confessional narrative.

The twelfth-century discovery of autobiography, moreover, did not bring an end to this authorial urge to insert one's own life into a biographical narrative. As we noted earlier, Adam of Eynsham instructs his readers through a vision that the Hugh of Lincoln they are meeting is not exactly Hugh himself, but instead is Hugh as Adam imagined him. Less allegorically, Walter Daniel tells his audience that the Ailred of Rievaulx they know – whether through the biography or through Ailred's own work – is very much a product of Walter's own labour, intellectual and physical. After describing Ailred's habits of speech, Walter Daniel concludes almost testily, 'But enough of this. His writings, preserved for posterity by the labour of my own hand, show quite well enough how he was wont to express himself.'[24] Readers of Joinville's *Life of Saint Louis*, written in the early fourteenth century, cannot help thinking that the book has been mistitled, since they learn far more about Joinville than about the work's titular subject.[25] Again and again the medieval writer imposes his own sense of self onto the text. Within every biography there is an autobiography surreptitiously making itself known.

The first straightforward autobiography in the Middle Ages, however, is Guibert of Nogent's. Guibert indicates his intention to take up Augustine's programme – the presentation of the evolution of the human heart – with the book's very first word: *Confiteor*. It is an oddly structured book, though the oddities are to some extent inherent in the autobiographical exercise itself. Book I describes the circumstances of Guibert's birth, his family life, his conversion to monasticism and his eventual discovery of true contemplation. Book II, the shortest of the three books, deals with his arrival at Nogent and the death of his mother, who often appears a shrill caricature of Augustine's mother, Monica. Book III tells of the politics surrounding the communal revolt in Laon, events in which Guibert participated but in which he was hardly a main character. Book I is no doubt the most confessional, in the Augustinian sense, in that it describes with remarkable intensity Guibert's own experience of conversion. The formal acceptance of Christianity is obviously not an issue for a twelfth-century European as it was for a fourth-century Roman. Guibert's embrace of monasticism, moreover, does not mark a true point of conversion, as it normally does in the traditional saint's *vita*. Rather, the conversion is a drawn-out process, which only reaches a still imperfect conclusion when Guibert attains a sense of peace with his life in his original monastery of Saint-Germer de Fly. Guibert then faced the same problem with which Augustine had to contend at the end of Book IX of his *Confessions*. Where does a writer of a life go after the apparent end of the story, conversion, has been reached, but when the actual end of the story, death, has yet to occur? Whereas Augustine turned inward and examined his own thoughts and sensory life with a still greater intensity, Guibert turned outward and looked at his world, particularly at the city of Laon. Whereas Augustine can achieve a sense of peace at the end of his *Confessions*, finishing the story of his life with an exegesis of the narrative of Creation, Guibert finds only chaos, the details of his life swallowed up by a world apparently bent on self-destruction. The structure of Guibert's memoirs are thus a perfect inverse of Augustine's, a turning outward after conversion and an openness to the world's cacophony, or perhaps an impotence before it.

The question we ought to ask about the *Monodies*, and which Guibert actually permits us to answer, is why did a relatively obscure monk in northern France take up an autobiographical programme, a literary exercise which had attracted almost no interest for 700 years? Is it simply because of a renewed interest in Augustine's *Confessions* in the twelfth century? We do know that Augustine's book did hold a special fascination for twelfth-century readers. Walter Daniel informs us that Ailred kept a copy of the *Confessions* alongside St John's Gospel (and a reliquary and a crucifix) inside his oratory as a special treasure.[26] But the discovery of a model is not by itself an explanation. Rather, there must exist within writers and readers an intellectual outlook which would make a book like the *Confessions* something that they would actually want to imitate. Guibert certainly knew of the *Confessions*, but he never refers to it directly.[27] His goal was not to parrot Augustine but to commit to parchment the details of his own life, no matter how extraneous or inconsequential many of them might seem. And his life *did* seem inconsequential to medieval readers. For while his memoirs circulate widely today, no single medieval exemplar of the text survives.

So what did inspire or enable Guibert – along with a handful of better-known later writers like Peter Abelard, Suger of Saint-Denis and Gerald of Wales – to set down the story of his life?[28] Guibert gives us the necessary clues in the memoirs themselves. As a young man, he writes, he learned from his mentor, Anselm of Bec, to use the Bible to speculate on the structure of the mind. Anselm taught him to think about the mind – how it was structured and how it functioned – and to look behind the tropological imagery of the Bible to find evidence for how these various parts interacted and gave shape to one another. It seems a harmless enough passage, useful as an insight into Anselm's teaching methods, but perhaps little else. The editor of the most famous English-language edition of Guibert's work goes so far as to suggest that Guibert is being disingenuous, and that in fact he never did follow Anselm's instruction.[29] It is, however, as I have argued elsewhere, the key to understanding all of Guibert's literary output.[30] Anselm, Guibert says, taught him to divide the mind into four parts: Intellect, Reason, Will and Affection. After their lessons, Guibert applied this model to the book of Genesis to see if he could find within it evidence for this idea. The exercise is not unlike writing a saint's *vita*, where the author finds a second narrative within the shadows of the events he describes, namely the narrative of the life of Christ. Guibert found within the shadows of Genesis a second narrative, a *mystica narratio* in his own words,[31] a narrative which told the story of the development of the mind.

The specifics of Anselm's lessons are not so important here. What is worth noting is that Anselm taught Guibert to think about (in the abstract) how the human mind develops – something equivalent to the evolution of the heart or to the stuff of autobiography. It is an interest which shows through in several places within Guibert's memoirs. His famous analysis of eleventh-century education, particularly his critique of corporal punishment, is one such example. 'Any person's nature,' he writes,

> let alone a child's, ends up being blunted if it has to submit to too much intellectual work . . . Alternately thinking about one thing, then another, we should then be able to come back to the one that our mind is most interested in, as if renewed by the recreation we had given ourselves.[32]

The debt which the thought owes to Anselm is clear from a passage in Eadmer's *Vita S. Anselmi*, where Anselm rebukes an unnamed abbot for his treatment of monastic oblates, who become useless and twisted due to too much confinement and too strict discipline.[33] In sum, the precondition for Guibert to write his *Monodies* was abstract speculation on the human condition, on human psychology, on what causes a person's character to develop in a particular way. The technical language is that of Intellect, Reason, Will and Affection. An example of practical application is a consideration of how to educate and discipline children. The result of such flights of theoretical speculation about the self is the first western autobiography in seven centuries.

At the heart of the language of Anselm and Guibert is the most traditional of Benedictine and monastic thought, which is to say the writings of Gregory the Great. Gregory's particular influence on both Anselm and Guibert is well established. Anselm was a student of Gregory's *Pastoral Care* and re-established the celebration of the ordination of Gregory as a Canterbury feast. Guibert began his education on Gregory's feast day and explicitly offered Gregory as a model for how to read the Bible.[34] Carol Straw has outlined brilliantly the patterns of thought to which serious readers of Gregory, like Anselm and Guibert, would have been exposed. The life of man is one 'riddled with polarities', which can both complement and repel one another. 'To find the world arranged in such clear and systematic oppositions is to acknowledge the divisions in human experience, yet also to awaken the hope and need to resolve such aching tensions and create a unified, harmonious whole, like life before the fall.'[35] An inability to attain perfection in this life is inevitable. It is 'the sacrifice of humility' that requires the contemplative saint to give up his heavenly insights and return always to the dangers of a carnal world.[36] The human mind can thus never escape the instabilities, tensions and oppositions of the world. Immediate setbacks upon advances are inevitable. Or as Guibert observes in the opening lines of his *Monodies*, 'I confess to your majesty, O God, the innumerable times I strayed from your paths, and the innumerable times you inspired me to return to you'.[37] What Guibert has done is to take the abstract ideas of Gregory about the back-and-forth motions of a human life and to apply them – notably, in a way that Gregory himself did not do in his *Vita S. Benedicti* – to the story of a life. The *mystica narratio* is not one of perpetual ascent, of a steady pilgrimage towards a higher state, but rather one of miscues, missteps, outright failures and very occasional advances. What Guibert has learned through Anselm is to apply these abstract, Gregorian ideas to a concrete problem – his own life. The birth of medieval autobiography, then, lies in the development and application of a medieval literary – or psychological, or tropological – theory. Twelfth-century writers like Guibert did not discover the individual or the self, but they did discover something as important: tools for thinking about and writing about the individual.

One can make a similar observation about the *Historia calamitatum*, the famous memoir-letter of Petter Abelard. Unlike Guibert, Abelard did have obvious and compelling reasons for telling his life story – namely to defend himself against several prominent critics and several powerful enemies. Abelard also wanted to leave a record of his famous relationship with Heloise that best suited his own memories of the affair and that best explained his own conduct.[38] But Abelard had one reason for writing about his life, one less obvious and one that makes him similar to Guibert: he too had engaged in abstract speculation about the shape of human consciousness. Abelard's most famous such exercise was his *Ethics*, or *Scito teipsum*, 'Know Yourself'. Questions of *intentio*, of the workings of the Will, were essential to Abelard's thought as much as to Guibert's. Abelard completed the *Ethics* in 1139, about a decade after the *Historia calamitatum*. But he based the treatise on several years of teaching and speculation, dating back at least until the 1120s.[39] Similar concerns shape both texts. In the words of John Marenbon, one of the purposes of the *Historia calamitatum* was for Abelard 'to make himself into an ethical example'.[40] Repeatedly throughout the *Historia*, Abelard similarly defends his *intentiones*, whether it was a question of why he attacked the validity of historical tradition at Saint-Denis, why he dedicated his oratory to the Paraclete rather than more conventionally to the Trinity or why he spent so much time in the company of nuns in later life. Even more obviously than Guibert, Abelard is a thinker who theorized about the self, and it was this speculation which gave him the tools to write about his own life in such a compelling fashion.

Subsequent medieval autobiographies and biographies would grow up from this intellectual base. A theoretical dialogue, most probably initiated by Anselm of Bec, though based on much older Augustinian and Gregorian ideas, started in northern France and gave to churchmen, and eventually to a wider community of writers and readers, the necessary intellectual tools to reconceptualize the meaning and the shape of human life. One might object that the monastic moral vocabulary of Anselm and Guibert, or the refined scholastic speculations of Abelard, provide too limited a basis for the reconceptualization of human psychology, and by themselves would have been unable to transform literature or conceptions of narratives of human life in the way I have suggested. A concern with emotion, with the complexity of personality, with interior life and motivation are characteristic of European society more generally. And undoubtedly one ought to consider survivals from popular literature as well as universal ecclesiastical practices, such as penance and confession, when thinking about how the autobiographical impulse might have spread.[41] On the other hand, it is well to remember how rapidly and how widely this very technical vocabulary could have travelled. Bernard of Clairvaux adopted the basic concepts of Anselm in his own treatises, and through him these writings spread throughout Cistercian houses and beyond.[42] Abelard's thought, despite the condemnations it received, would give direction to university speculation and no doubt had its impact on what Abelard's enemy, Bernard, wrote and thought as well.[43] When Chrétien de Troyes imagined Lancelot's hesitation before leaping into the cart, and then shaped the remainder of his story's action around the consequences of that moment's doubt, he certainly did so under the influence of these intellectual currents.[44] The preaching orders in the thirteenth century further popularized these ideas. Debates about mortal sin at the court of Louis IX, overseen by friars educated at the university, most certainly affected the way Joinville presents his own actions and motivations while on Crusade.[45] These preachers and confessors further circulated the ideas to audiences beyond the schools, the monasteries and the aristocracy. When Margery Kempe, concerned about the origins of her visions, consulted with a friar about them, he informed her that Christ himself had placed 'these condicyons eythyr in your wyl or in your affeccyon er ellys in bothyn . . .'.[46] One cannot help but feel surprise at finding, in the fifteenth century and in English, half of the psychological vocabulary of

Guibert of Nogent. Whether Margery understood these words in the same way as did Guibert or Anselm or Bernard is doubtful. But her own autobiography does demonstrate that the core theoretical ideas were very much in circulation. Amateur observers today regularly diagnose an acquaintance with an Oedipal complex or an identity crisis or an inferiority complex, without understanding the term's intellectual geneology or, for that matter, what it actually means. But the essential concepts, however far removed from how students of psychology understand them, have given shape to the ways in which ordinary people interpret and explain the human condition. So did the autobiographical thought which took root in northern France in the late eleventh and early twelfth centuries cause a wide-ranging reconceptualization of what a life is and how the story of a life ought to be told.

As the above fomula suggests, it is not surprising that the same theoretical concerns which helped to create medieval autobiography would also reshape medieval biography. It is less surprising still that one of the first biographical texts to demonstrate such concerns would be Eadmer's *Vita S. Anselmi*, produced as it was within the very intellectual circles in which Guibert and Abelard moved. Many aspects of this text make it remarkable. Sir Richard Southern, for example, has called attention to the unusual care with which Eadmer records Anselm's conversation in ways that seem to provide a direct window onto the archbishop's soul.[47] An equally important and related aspect of the text is the psychological detail which Eadmer sprinkles into these conversations. When Anselm is debating whether to become a monk and, if so, which monastery he ought to join, he considers two options – Cluny and Bec. The burden of the Cluniac liturgy, he realized, would allow him little time for the writing and speculation for which he excelled. At Bec, on the other hand, he feared he would always be in the shadow of Lanfranc, the monastery's most famous scholar. Though he eventually decided on Bec, his intentions and desires continued to fluctuate.[48] A few years later, when he conceived the theological project which was to become the *Proslogion* – whether it was possible to demonstrate the existence of God by a simple, brief argument – he found the project so absorbing that he was unable to eat, drink, sleep, or concentrate on the proper performance of liturgy. Eventually he began to wonder if the idea were a demonic temptation, an impossible task destined to bring him more harm than good. But the more he tried to abandon the inquiry, the more it haunted him.[49] This sense of doubt, of anxiety, of regular self-questioning about one's intentions – all of this places Anselm's character, and Eadmer's version of his life, firmly within the traditions of Guibert and Abelard. And even though Anselm seemed a likelier fit for a traditional saint's *vita* than either Guibert or Abelard, what gives shape to Eadmer's account is not a steady pilgrimage towards unity with God. It is instead the uncertainties, the fluctuations of the Will, the moments of self-doubt and the passages of conversations overheard – all of which together reveal the tensions which reverberate within the human heart.

Equally remarkable, however, are the moments when Eadmer shows his readers the tensions and uncertainties within his own heart. We see him disappointed when he is able to break off only a small fragment of bone from the relics of Saint Prisca, though Anselm reassures him that the saint will give him just as great a reward for taking care of this small bit of her physical remains as she would have done had Eadmer got possession of the entire body.[50] We sense his excitement at meeting Anselm as a youth and we hear his anger at his fellow monks whose grumblings have begun to detract from Anselm's legacy. Above all, we hear Eadmer dissecting his own intentions and his own possible culpability at disobeying Anselm's command to destroy the manuscript of the *Vita S. Anselmi*.[51] This last act of disobedience causes Eadmer to ask his readers – or at least those who have enjoyed his narrative – to intercede with God, and probably with Anselm, too, on his behalf. Eadmer perhaps tells us more about his insecurities about his relationship with Anselm than he intends here when he writes that he will never forget how Anselm himself had promised to intercede for his disciple's soul: 'He said certainly he would willingly and gladly do this, only let me take care not to make myself too heavy for him.'[52] The uneasiness which Anselm felt at Eadmer's project perhaps stems from Anselm's own understanding of the biographical art. The Anselm which Eadmer would give us would not be the one Anselm wished us to have; for better or worse, it would be Eadmer's Anselm and not the sense of self which Anselm had sought to create in his theology and his letter collections.[53]

All of these characteristics – a tendency towards free-flowing narrative, a habit of psychological speculation and intimate autobiographical expressions by the biographer himself – appear readily in later medieval biographical writing. Consider again Adam of Eynsham, the constant companion of Hugh of Lincoln, so much so that he stayed by his bishop's bedside and attempted to decipher the words the bishop spoke as he slept.[54] In order to express the necessary sense of wonder at Hugh's generosity towards lepers (perhaps essential, since visits to leprosaria were becoming somewhat of a hagiographical commonplace), Adam opens up to his readers his own mind, obviously in the expectation that his audience will identify with him.

> Have pity, sweet Jesus, on the unhappy soul of the narrator! I cannot conceal, would it were concealed from your vengeance, how much I shuddered not merely to touch but even to behold those swollen and livid, diseased and deformed faces with the eyes either distorted or hollowed out and the lips eaten away![55]

Adam, like Eadmer, does not just reveal his own thoughts and reactions but also speculates for us about Hugh's psychological state. For instance, he describes how

Hugh must have felt when he decided to leave his first monastery – and thus to break his vow – in order to take up the more rigorous calling of the Carthusians: 'Merciful Heavens! What a terrible conflict between two loves raged in the heart of thy servant at this appeal! What a horrible and awful dilemma it was!'[56] To break a monastic vow in order to join a stricter order was considered acceptable in twelfth-century Europe, but it did no doubt trigger crises of conscience such as the one Adam describes. Still, that fact should not prevent us from noting that Adam, in fact, had no evidence for his portait, since he concludes the story: 'Your servant frequently asked you whether sometimes, as might be expected, your broken oath caused you the least misgiving, and you always replied "I never felt the least misgiving but rather heartfelt joy whenever I remembered an act so profitable to myself."'[57] Adam admits when he describes his vision of the saint's death, he does not give us the real Hugh. He gives us instead Hugh as he preferred to understand him.

At the beginning of this chapter we noted that medieval biography and autobiography have stood largely impervious to modern theoretical speculation. In focusing on the theoretical approaches which medieval writers themselves brought to their tasks, I may have done little to alter this situation. A few general rules and observations, however, have suggested themselves. First, all medieval biography is, to a degree, autobiographical. This is not to suggest, in a defeatist fashion, that we can never know the subject of the text but only a distorted image of the author or of the author's narrative voice. It is simply to say that we as historians and as readers of medieval Lives must learn to recognize this authorial voice and the ways in which it gives shape to what we read. It asserts itself in three major ways. In the most mundane sense, biographers often feel the urge to insert themselves into the text, to tell a story about themselves or their connection to the subject, whether they knew that subject directly, as Adam knew Hugh, or more distantly, as Osbern had experienced Dunstan through a pilgrimage to his cell. Second, an author of a biography – whether a saint's *vita* or a ruler's panegyric or a life story which appears within the context of a more general history – has a particular intellectual or political agenda. We must take into account the circumstances in which a text was written and the goals which an author had for his story before we seek to interpret the events it purports to describe. And finally, biographers have preconceptions about how a life ought to be led, preconceptions which grow partly out of theoretical concerns and partly out of their own experiences. The Life which they present is a life designed to demonstrate those ideals in action.

As a second general rule, and as an inverse of the first, all autobiographical writing is, to an extent, biographical. It is obvious enough that when biographers write about someone's life, they do so with a theoretical apparatus in the back or in the forefront of their minds. A life gains meaning not just from the actions which occur within it, but also from the ways in which it conforms to, illustrates or helps to refine a theory about society, about psychology or about any other subject worthy of speculation. The same is true for autobiographers. Guibert chose to write, and was able to write, his memoirs because Anselm had taught him to think about the mind and how it develops. His *Monodies* were the product of his theories. The same is true for Abelard. He could write the *Historia calamitatum* because he had in his capacity as a scholastic theoretician dissected, in all its complexity, what an Intention actually is. When autobiographers seek to define their particular selves, they do so with generalized constructs which they have fashioned within their particular intellectual worlds.

Both these points lead to a broader conclusion about interpretations of the new direction taken by European culture around the year 1100. Other commentators have spoken of a discovery of the individual or a twelfth-century renaissance, concepts which the character of twelfth-century texts readily supports. We ought not to view these changes as a sudden discovery of the self by twelfth-century men or women, and certainly we ought not to see it as evidence for a breakdown in group identities, which are a constant factor in medieval history. Rather, writers in the later eleventh century began to formulate new concepts with which to analyse the self. The roots of this otherwise elusive historical shift are thus textual and intellectual.[58] Anselm, Guibert, Abelard, Hugh of Lincoln and countless others do not come to life with such sophistication and complexity because their century *discovered* the individual; rather, their century's biographers discovered how to *think about* and then to *write about* the individual.

But we ought to note that medieval biographers shared one other bond with their subjects, one that transcends whatever intellectual tools they brought to their projects. Let us return to Adam of Eynsham's vision on the eve of Hugh's death, when Adam easily picked up the enormous tree trunk, representative of his own memory of Hugh, and when he saw how the leaves and branches had scattered from the tree, symbols of the pages of the book he would write. Medieval readers, as we know, lived comfortably with a multiplicty of symbolic meanings and Adam did indeed give his vision one other interpretation. The trunk was Hugh's corpse. 'Moreover, the leaves which he saw had fallen from the tree might symbolize the hairs round his tonsure and beard and his finger and toe nails, which the man beloved of God had ordered to be shaved and pared the evening after his death.'[59] For Adam was not simply Hugh's biographer, but also his embalmer. And it is striking how common this pattern is for the writers discussed here. Eadmer embalmed Anselm, rejoicing as the scarce oil miraculously multiplied and permitted him to anoint his master's entire body. Walter Daniel kissed the feet of his lifeless master, Ailred of Rievaulx, and then argued with his brethren over the best way to treat the body. Another twelfth-century biographer, Reginald of Durham, describes in such detail the preparation of Godric of Finchale's body – including the sudden flow of blood when the monks cut too deeply into Godric's toenails – that he surely was involved in the embalming process, too. The biography was perhaps the final act of preservation, the life and legacy of a beloved master set in order to survive the world's decay until Judgement Day. The laborious process of writing and of publication in the Middle Ages – letters cut first onto wax and then transferred onto animal skins – was a much more physical one than is the case for modern writers. The analogy to embalming would have been far more obvious. The tree trunk is a sign that Hugh will die. The tree trunk is Adam's memory of Hugh – not Hugh himself, but the basis for the book through which we know him. The tree trunk is Hugh himself, his body cradled lovingly by Adam of Eynsham in the manner of a *pietà*. This fluidity of symbolism is yet another example of the flexibility of the medieval mindset. It is also a tacit recognition by Adam of the fineness of imagination that separates Hugh's life from his memory, his biography from his biographer.

Guide to further reading

Bynum, Caroline Walker, 'Did the Twelfth Century Discover the Individual?', in *Jesus as Mother: Studies in the Spirituality of the High Middle Ages* (Berkeley, CA, 1982), pp. 82–109.

Clanchy, M.T., *Abelard: A Medieval Life* (Oxford, 1997).

Heffernan, Thomas J., *Sacred Biography: Saints and their Biographers in the Middle Ages* (Oxford, 1988).

Mews, Constant J., *The Lost Love Letters of Heloise and Abelard: Perceptions of Dialogue in Twelfth-Century France* (New York, 1999).

Misch, Georg, *A History of Autobiography in Antiquity*, trans. E.W. Dickes (2 vols, Cambridge, MA, 1950).

Morris, Colin, *The Discovery of the Individual, 1050–1200* (London, 1972).

O'Connell, Robert J., *St. Augustine's Confessions: The Odyssey of Soul* (New York, 1989).

Olney, James, *Metaphors of Self: the Meaning of Autobiography* (Princeton, NJ, 1972).

Rubenstein, Jay, *Guibert of Nogent: Portrait of a Medieval Mind* (New York, 2002).

Southern, R.W., *Saint Anselm and His Biographer* (Cambridge, 1963).

Notes

1   The literature on autobiography and its relationship to modern notions of the 'self' (as opposed to medieval notions) is enormous. Some notable works include James Olney, *Metaphors of Self: The Meaning of Autobiography* (Princeton, NJ, 1972), and James Olney (ed.), *Autobiography: Essays Theoretical and Critical* (Princeton, NJ, 1980); Robert Folkenflik (ed.), *The Culture of Autobiography: Constructions of Self-Representation* (Stanford, CA, 1993); and Kathleen Ashley, Leigh Gilmore and Gerald Peters (eds), *Autobiography and Postmodernism* (Amherst, MA, 1994). A recent, straightforward presentation of the thesis of autobiography as a product of modernity is Michael Mascuch, *Origins of the Individualist Self: Autobiography and Self-Identity in England, 1591–1791* (Stanford, CA, 1996). The major exception to the rule of associating autobiography with modernity is Georg Misch, *Geschichte der Autobiographie* (4 vols, Leipzig, 1907–69), which examines autobiographical texts from antiquity through the Middle Ages. Any explicitly medieval focus of the problem of autobiography must, inevitably, take as a central figure St Augustine, whose general influence is considered in Pierre Courcelle, *Les 'Confessions' de saint Augustin dans la tradition littéraire. Antécédents et postérité* (Paris, 1963). I have attempted my own analysis of the subject with an emphasis on Guibert of Nogent: Jay Rubenstein, *Guibert of Nogent, Portrait of a Medieval Mind* (New York, 2002), pp. 61–82.

2   Colin Morris, *The Discovery of the Individual* (London, 1972). The literature surrounding this question and the related 'twelfth-century renaissance' is, again, enormous. See the essays collected in Giles Constable and Robert Benson (eds), *Renaissance and Renewal in the Twelfth Century* (Cambridge, MA, 1982); the recent re-evaluation of the problem by R.N. Swanson, *The Twelfth-Century Renaissance* (Manchester, 1999); and, most recently, C. Stephen Jaeger, 'Pessimism in the Twelfth-Century Renaissance', *Speculum* 78 (2003), pp. 1151–83.

3   Adam of Eynsham, *Magna Vita Sancti Hugonis*, eds Decima L. Douie and Hugh Farmer (2 vols, London, 1961, 1962), 2, pp. 210–11.

4   Perhaps the most eloquent statement of the thesis that all writing is autobiography is Olney, *Metaphors of Self*.

5   Adam gives the years of his service in the prologue to Book 2 of the *Vita*: Adam of Eynsham, *Vita S. Hugonis*, vol. I, pp. 45–6.

6   Caroline Walker Bynum writes usefully on this topic, emphasizing the importance of group identities in defining the self in the twelfth century as opposed to a 'renaissance' or a 'discovery of the individual': 'Did the Twelfth Century Discover the Individual?', in *Jesus as Mother: Studies in the Spirituality of the High Middle Ages* (Berkeley, CA, 1982), pp. 82–109.

7   Benedicta Ward, *The Venerable Bede* (London, 1990), pp. 114–16.

8   Though commentators have detected literary and theological models in Gregory's work, too; among them: Walter Goffart, *The Narrators of Barbarian History (550–800): Jordannes, Gregory of Tours, Bede, and Paul the Deacon* (Princeton, NJ, 1988) and Martin Heinzelmann, *Gregory of Tours: History and Society in the Sixth Century*, trans. Christopher Carroll (Cambridge, 2001).

9   And at least in the early stages of composition, that is what Adam believed he was doing: 'Ut autem cepte narrationis ordinem seriatime prosequamur . . .' (Adam of Eynsham *Vita S. Hugonis* 1, p. 12).

10 Eadmer, *Vita S. Anselmi*, ed. and trans. R.W. Southern (Oxford, 1962), p. 2.

11 On the ability of medieval thinkers to accept comfortably multiple interpretations, see especially Philippe Buc, *L'ambiguïté du Livre: prince, pouvoir, et peuple dans les commentaire de la bible au moyen âge* (Paris, 1994).

12 Three excellent introductions to this topic would be: Henri de Lubac, *Exégèse médiévale: les quatre sens de l'Écriture* (4 vols; Paris, 1954–64), Jean Danielou, *From Shadows to Reality* (London, 1960), and Beryl Smalley, *The Study of the Bible in the Middle Ages* (University of Notre Dame, IN, 1964).

13 Adam of Eynsham, *Vita S. Hugonis* 1, p. 79, referencing Genesis 19:17–23 and Exodus 20:21.

14 The classic introduction to hagiography is Hippolyte Delehaye, *The Legends of the Saints*, trans. Donald Attwater (New York, 1962). See also Thomas J. Heffernan, *Saints and their Biographers in the Middle Ages* (Oxford, 1988).

15 The Lives of St Martin, by Sulpicius Severus, and St Benedict, by Gregory the Great, are most accessible in *Early Christian Lives*, trans. Caroline White (London, 1998), pp. 131–59 and 163–204, respectively. For Guthlac: Felix, *Life of Saint Guthlac*, ed. and trans. Bertram Colgrave (Cambridge, 1956). Whenever possible, saints' Lives such as these, available in multiple editions and translations, will be cited according to chapter number as well as page number.

16 The various Lives of Dunstan, by 'B', Adelard, Osbern, Eadmer and William of Malmesbury, are collected in W. Stubbs (ed.), *Memorials of St Dunstan* (London, 1874). Partial translations appear in *English Historical Documents* 1, *500–1042*, ed. Dorothy Whitelock (London, 1956). The classic account of Wilfrid's stormy political career is that of Eddius Stephanus, *The Life of Bishop Wilfrid*, ed. and trans. Bertram Colgrave (Cambridge, 1927). Anselm's biographer, Eadmer, wrote a life of Wilfrid as well, attempting the impossible task of making it more 'Canterbury friendly', and it has recently been published as: Eadmer, *Vita Sancti Wilfridi*, eds Bernard J. Muir and Andrew J. Turner (Exeter, 1998).

17 Peter Brown, *Augustine of Hippo* (London, 1967), p. 28. The phrase
    and underlying concept are similar to Carolyn A. Barros's
    Autobiography: Narrative of Transformation (Ann Arbor, MI, 1998). See
    also Robert J. O'Connell, *St. Augustine's Confessions: The Odyssey of
    Soul* (New York, 1989) and Thomas Renna, 'Augustinian
    Autobiography: Medieval and Modern', *Augustinian Studies* 11 (1980),
    pp. 197–203.
18 Augustine, *Confessions* X, 30 (cited by book and chapter, but using the
    translation in this edition: trans. F.J. Sheed (Indianapolis, 1942, 1993)).
19 The quotation is from Guibert's earliest writing, his *Opusculum de
    virginitate*, published in Migne (ed.), *Patrologia Latina* 156, col. 581A.
20 Gregory the Great, *Vita S. Benedicti*, 2. One might refer to the transla-
    tion by Caroline White in *Early Christian Lives*, pp. 268–9, as cited
    above.
21 Osbern, *Vita S. Dunstani*, §13.
22 R.W. Southern, *Saint Anselm and His Biographer* (Cambridge, 1963),
    p. 231, and Jay Rubenstein, 'The Life and Writings of Osbern of
    Canterbury', in *Canterbury and the Norman Conquest*, eds Richard Eales
    and Richard Sharpe (London, 1995), pp. 29–30.
23 Jay Rubenstein, 'Liturgy against History: The Competing Visions
    of Lanfranc and Eadmer of Canterbury', *Speculum* 74 (1999),
    pp. 279–309.
24 Walter Daniel, *The Life of Ailred of Rievaulx*, ed. and trans. Maurice
    Powicke (London, 1950), p. 18.
25 Joinville's *Life of Saint Louis* is most readily accessible in Joinville and
    Villehardouin, *Chronicles of the Crusades*, trans. M.R.B. Shaw
    (London, 1963).
26 Walter Daniel, *Vita S. Ailredi*, p. 51.
27 Courcelle, *Les Confessions*, pp. 272–5.
28 Good recent overviews of all these (very different) autobiographers and
    their worlds have been published: M.T. Clanchy, *Abelard: A Medieval
    Life* (Oxford, 1997); Lindy Grant, *Abbot Suger of St-Denis: Church and
    State in Early Twelfth-Century France* (New York, 1988); and Robert
    Bartlett, *Gerald of Wales, 1146–1223* (Oxford, 1982). Bartlett's discus-
    sion of Gerald's conflicted identity due to his 'Welshness' is especially
    provocative.
29 John Benton, *Self and Society in Medieval France* (New York, 1970),
    p. 91 n. 16.
30 Rubenstein, *Guibert of Nogent: Portrait*, pp. 38–60.
31 Guibert of Nogent, *Moralia Geneseos*, in *Patrologia Latina* 156,
    col. 64B.
32 Guibert of Nogent, *Monodies* 1, 15 (following the translation by Paul J.
    Archambault, *A Monk's Confession* (College Station, PA, 1996), p. 17).
33 Eadmer, *Vita S. Anselmi*, pp. 37–9.

34 On Anselm and Gregory, R.W. Southern, *Saint Anselm, a Portrait in a Landscape* (Cambridge, 1990), pp. 234–5; on Guibert's education, *Monodies* 1, 4; he identifies Gregory as a model in his preacher's manual, *Liber quo ordine sermo fieri debeat, Corpus Christianorum continuatio medievalis* 127, p. 55.

35 Carol Straw, *Gregory the Great: Perfection in Imperfection* (Berkeley, CA, 1988), p. 52; see also her diagram on p. 54.

36 Straw, *Gregory the Great*, p. 189.

37 Guibert, *Monodies* 1, 1 (trans. Archambault, p. 3).

38 This point demonstrated most recently by Constant Mews, *The Lost Love Letters of Heloise and Abelard: Perceptions of Dialogue in Twelfth-Century France* (New York, 1999). Mews' book is in many respects a model for how to uncover the abstract constructs which govern the creation of an autobiographical text.

39 On the dates and the background of the *Ethics*, see John Marenbon, *The Philosophy of Peter Abelard* (Cambridge, 1997), pp. 66–7.

40 Marenbon, *Philosophy of Peter Abelard*, p. 320.

41 Mary C. Mansfield, *The Humiliation of Sinners: Public Penance in Thirteenth-Century France* (Ithaca, NY, 1995).

42 Jean Leclercq demonstrates the intellectual and geographic connections between Cistercian and chivalric imagery in his study *L'amour vue par les moines au XIIe siècle* (Paris, 1983).

43 Clanchy, *Abelard*, esp. pp. 25–40 and 288–325. See also D.E. Luscombe, *The School of Peter Abelard: The Influence of Abelard's Thought in the Early Scholastic Period* (Cambridge, 1969) and A. Victor Murray, *Abelard and St. Bernard: A Study in Twelfth-Century 'Modernism'* (Manchester, 1967).

44 Chrétien's story of Lancelot appears in a variety of editions. The scene with the cart occurs at lines 360 ff. See, for example, Chrétien de Troyes, *Arthurian Romances*, trans. William B. Kibler (London, 1991) pp. 211–12.

45 The debate sequence occurs in the first chapter of Joinville's *Life of Saint Louis*, pp. 168–70 of the Shaw translation.

46 *The Book of Margery Kempe*, ed. Lynn Staley (Kalamazoo, 1996), §18, p. 53.

47 Southern, *Saint Anselm: A Portrait*, pp. 422–8.

48 Eadmer, *Vita S. Anselmi* 1, 5–6, pp. 8–11.

49 Eadmer, *Vita S. Anselmi* 1, 19, pp. 28–31.

50 Eadmer, *Vita S. Anselmi* 2, 45, pp. 132–4.

51 For the above incidents, see Southern's edition of the *Vita S. Anselmi*, pp. 132–4, 50, 170 and 150–1.

52 Southern's translation, *Vita S. Anselmi*, p. 151.

53 Sally Vaughn offers a similar suggestion in her article, 'Eadmer's Historia Novorum: A Reinterpretation', *Anglo-Norman Studies: The Proceedings*

of the 10th Battle Conference, Caen Normandy, ed. R. Allen Brown
(Bury St Edmunds, 1988) pp. 269–89.

54 Adam of Eynsham, *Vita S. Hugonis* 1, p. 75.

55 Adam of Eynsham, *Vita S. Hugonis* 2, p. 13.

56 Adam of Eynsham, *Vita S. Hugonis* 1, p. 26.

57 Adam of Eynsham, *Vita S. Hugonis* 1, p. 27.

58 The key contention here – that in order to understand medieval narratives one must be aware of authorial programmes and of authentically
medieval intellectual systems – is similar to the one presented about the
place of ritual in early medieval historical narrative by Philippe Buc, *The
Dangers of Ritual* (Princeton, NJ, 2001).

59 Adam of Eynsham, *Vita S. Hugonis* 2, p. 211.

3

The hidden self: psychoanalysis and the textual unconscious

Nancy Partner

One of the greatest impediments to recognizing the depth, complexity and individuality of the people who lived during the immense span of historical time we categorize as 'medieval' is that 'medieval' has come to mean the opposite of those qualities, at least as regards persons. Medieval culture, in terms of its art, literature and theology, has long been acknowledged as sophisticatedly complex and emotionally dense, and the term 'scholastic', while not exactly a compliment, at least suggests a demanding intellectual discipline, abstract and rigorous. But somehow this collective cultural achievement is oddly disconnected from any idea of medieval persons of equivalent individual complexity. The general absence from medieval evidence of the sorts of writing that document the interior life – diaries, personal letters, intimate memoirs and revelations of private life – and a prevalence of didactic genres (ranging from epic to sermons) which stress conformity with religious and social norms encourage the notion that in some way the pre-modern era of history was populated with pre-individuals. This is probably a secondary effect of literary production in a society where literacy was largely the special skill of a professional clerical elite, but the impulse to generalize from sector to society is nearly irresistible. Vast stretches of medieval history do not bring to mind an imaginary dramatis personae fraught with interior tension and personality like the vivid personages of the Elizabethan court or Samuel Pepys confiding his sex life to his diary in specially invented code, Samuel Johnson caring about his cat in eighteenth-century London. The 'self', a highly conceptual, if ill-defined, term we use as a container for the elements of individuality that differentiate persons from the mass – self-consciousness, desires, conflicts, aware interiority layered over the iceberg depths of unconscious mind, all charged with the positive value of agency – just does not naturally attach to 'medieval' persons. We are too aware that medieval documents fail to speak of the private plans, the guarded responses of 'my' life, 'my' being, that fill out and fill up our sense of an individuated self, and it is easy to let the very term 'medieval' blanket and muffle individual personality and selfhood.

But these ideas (which oddly don't seem to apply to Greek and Roman antiquity, periods hardly richer in diaries and personal confidences) are grossly pejorative and this prejudice has a history, and a fairly recent one in its modern enunciation. The dramatic contrast between typical medieval persons and the new man of the Renaissance drawn by Jacob Burckhardt in his *Civilization of the Renaissance in Italy*, first published in 1860 (and perennially in print, in translation, with a new edition as recently as 2002), shaped ideas of medieval and Renaissance life with a few brilliant exaggerated strokes and has persisted to become an indelible cliché.

> In the Middle Ages both sides of human consciousness – that which was turned within as that which was turned without – lay dreaming or half awake beneath a common veil. The veil was woven of faith, illusion, and childish prepossession, through which the world and history were seen clad in strange hues. Man was conscious of himself only as member of a race, people, party, family, or corporation – only through some general category. In Italy this veil first melted into air; an *objective* treatment and consideration of the State and of all the things of this world became possible. The *subjective* side at the same time asserted itself with corresponding emphasis; man became a spiritual *individual*, and recognized himself as such.[1]

Such ideas, albeit less romantically expressed, have become entrenched in popular ideas of 'medievalness', mildly repudiated but not replaced by specialist scholars. Clichés, one notes with resignation, are the most tenacious cultural forms. Writing the introduction to a paperback edition of *Civilization of the Renaissance* in 1958, Benjamin Nelson and Charles Trinkaus note about this stereotype: 'Rarely has an historical work had so persistent an influence.' They also acknowledge that this strand of influence 'never fails to distress professional students of the Middle Ages', and plead that Burckhardt never anticipated the effect of the few sentences he devoted to medieval minds, such as, 'no such conscious division of experience into subjective and objective existed in the Middle Ages'.[2] Modern medievalists emphatically do not subscribe to these derogatory commonplaces – ascribing the mental life of early infancy to 'the Middle Ages' – but it remains difficult to say what exactly has replaced them. In 1990, the Chaucer scholar, Lee Patterson, pointed to the dragging persistence of Burckhardt's notion of medieval (non-)selfhood into current scholarship where it fitted the convenience of early modern scholars who could define modernity against undeveloped medieval minds.[3] But erasing a set of dated clichés from current academic vocabulary does not, all by itself, work *in the positive* to bring complex-feeling medieval persons into the imagined historical field in the same way that we can (and do) 'think with' rich personality types like Martin Luther, Petrarch, Benvenuto Cellini, even Romeo and Juliet. Thomas Acquinas, in typical sorry contrast, is an abstract theology with a man's name attached, different by barely a degree from the Anonymous of Rouen – no-name author of an eleventh-century pamphlet collection.

What helps, I think, is to press harder than we usually do on the concept of the self operating silently here – not everything can be blamed on inadequate evidence. The idea of the self underlying Burckhardt's celebratory portrait of a certain Renaissance style is a great deal more specific and limited than is usually recognized. It is clear from many passages – in fact, all the passages of his Part II, 'The Development of the Individual' – that his great subject is not selfhood (a condition of human mental life) but *individualism* (a style and set of values), a quite particular style of floridly performed and publicly rewarded self-cultivation. Burckhardt focused narrowly on the development of talents that show vividly in competitive social life, especially among ambitious men vying for the attention of patrons and the politically powerful in court society of the fifteenth and sixteenth centuries – a historically local style. But there lingers a common and unexamined assumption that 'having' a self, or evincing the existence of the mental and emotional organization specified as selfhood, the self we feel as our own identity, necessarily involves adopting one assertive style of individuality, even the set of values and goals we associate with the individualism which grounds western liberal modernity. Pre-modern societies that endorsed or gave authoritative approval to high degrees of conformity in thought and behaviour (as in monasteries), cultivated literary conventions of modesty and deference to authority and socialized their young to an ethos of service and hierarchy, are vaguely thought to have suppressed the self. Lacking the life details and personal revelations that keep the proof of inner depths and tensions before our historical imagination, it is too easy to let medieval people sink down into a shallow bas-relief of 'medievalness', defined by the moralizing conformist elements of the dominant literate culture. To recognize and think clearly about the self in its medieval guises, it is important to make a crisp distinction between the individual self and the particular styles of individual self-display encouraged or tolerated, or constrained and condemned, within a specific cultural surround.

Even the cultivated styles of 'selflessness' associated with religious communities, mystical and ascetic practices, or acquiescence in gender and social hierarchies, are still the actions of a human mind, the vector result of conscious and non-conscious pressures and compromises, aimed at certain ends for certain reasons, some known, some not. The surface of behaviour never tells the whole story, any more than the literal level of a text. Self-aware interior life was certainly acknowledged in ancient and medieval culture, although veiled (to us) in different metaphorical language alluding to pre-modern psychology. The heart, not the head (mind/brain), was the characteristic medieval metaphoric location of feelings, memory, knowledge, the centre of individuality; and the book of the heart was one common figure of speech which linked interior life to language, the heart itself *as* a book. 'The book of the heart is a quintessentially medieval trope . . . to figure a book as a heart was to equate textuality with subjectivity, since the heart was central to medieval psychology'. From Augustine's *Confessions* to the advent of print technology, the personal writing inscribing memories and feelings, usually on religious subjects, cast in a spiritual vocabulary in the 'book of the heart', was understood to refer to interior life, different for each person. 'During the Middle Ages these now-dead metaphors combined a vivid imagery of the book with a pectoral psychology that saw the heart as the psychosomatic center of the human being.'[4] Modern scholarship, most notably by Mary Carruthers, investigating the ways that medieval people were aware of and found a language for their own mental processes, has given us acute and sensitive studies of the arts of memory and the craft of thought; these studies teach us the terms of medieval self-understanding (adopted from classical rhetoric and mnemonic arts) and provide a crucial bridge to modern theories of the mind–language connection.[5] Pre-modern psychologies grounded in Aristotle and Augustine are themselves historical subjects for scholarly recovery and understanding, but current interpretive analysis of historical persons has to proceed from widely acceptable concepts, not themselves embedded as part of the historical context.

The need for a depth psychology in medieval studies has been emphasized by the great modern expansion of the field beyond the traditional areas of political and constitutional history. By turning to social history, women's history, the family, history of childhood, sexuality and gender, medievalists have introduced the interiorized self in all its complexity into the historical field. If we insist on wanting to know how men and women comported themselves in their marriages, why they thought prostitution should be regulated by law, what they thought about the gender norms that regulated their behaviour and how they acted in spite of their thoughts, what feelings meshed with their religious beliefs, how women confronted the myriad restrictions and constraints that defined their lives, then we have to decide, in more than a slovenly taken-for-granted way, just what kind of people we think we are talking about. This understanding proceeds from acknowledging medieval men and women as essentially *like ourselves*, of the same species at the same moment of development in evolutionary time, personalities formed at a deep level through the same developmental processes, as minds with the same emotional/rational structure confronting the world, however distractingly different their language, ideals and fervent beliefs. Such recognition of human commonality is nothing like the 'essentialism' of reductive stereotypes rightly condemned throughout the historical discipline.

Everyone's unconscious: the acts of the mind

The discipline of psychoanalysis, with its coherent structure of explanatory concepts, is our intellectual instrument for recognizing the human psyche over historical time and across cultures. Begun and established by the work of Sigmund Freud, refined by over a century of continuing clinical and theoretical work, we now have a richly developed and lucid theory for many interpretive purposes. Regardless of whether anyone ostensibly 'does not accept' psychoanalysis, its basic language and concepts permeate our culture and are invoked daily in every conceivable context. People refer matter-of-factly to repression, projection, stereotyping, fantasy, being 'in denial', the long-term influence of childhood events, the family romance, Oedipal conflict and dream meanings, often without realizing the source of these powerful ideas. The strength of psychoanalytic theory in offering coherent and intelligible accounts of mental life beyond conscious intentions is such that even scientific fields such as neuroscience that earlier sought to replace it entirely are now forming connections with psychoanalysis as a complementary theory. Historians who hesitate over adopting what they have vaguely heard to be an outmoded or disproved theory may be reassured to learn of the newest positive connections between psychoanalysis and neuroscience: as a useful summary article in *Scientific American* puts it, 'neuroscientists are finding that their biological descriptions of the brain may fit together best when integrated by psychological theories Freud sketched a century ago'.[6] For our signal purpose of reading historical evidence, cognitive science cannot replace or supersede a depth psychology because it cannot do the interpretive work that analytic concepts bring to memory, fantasy, language, all symbolic structures. For students of history it is especially fascinating to see the ways in which psychoanalysis stands as a bridging connector point between scientific laboratory-based studies of brain function and pre-modern philosophy and theology addressed to the mind, language and hidden realms of meaning.

As an exploration of the human condition, and particularly the deep complexity of human minds, psychoanalysis stands in a long tradition traced by philosopher and analyst Jonathan Lear, through Plato, Saint Augustine, Shakespeare and into modern thought: 'What holds this tradition together is its insistence that there are significant meanings for human well-being which are obscured from immediate awareness'.[7] Psychoanalysis begins with the premise that much of mental activity is unconscious – that is, not immediately available to awareness, introspection or observation from outside. This basic perception of the depths of the mind/soul was present to pre-modern culture, but unlike ancient and medieval theories for locating hidden meaning, Freud's work located the sources of meaning within the human world without reaching for divine intervention or requiring a specifically religious world-view. In Lear's words:

Freud's achievement, from this perspective, is to locate these meanings fully inside the human world. Humans make meaning, for themselves and for others, of which they have no direct or immediate awareness. People make more meaning than they know what to do with. This is what Freud meant by the unconscious.[8]

Analytic theory offers historians the interpretive techniques and vocabulary for moving from manifest to latent levels of meaning without demanding implicit acceptance of an ideology or divinity.

Medievalists surely have come to agree with Sir Richard Southern's prescient remarks, in 1961, that history's proper focus extends beyond institutions and politics to 'the study of the thoughts and visions, moods and emotions and devotions of articulate people. These are the valuable deposit of the past.'[9] The chief obstacle apparently posed by psychoanalysis for historical uses, and one that must seem formidable even to historians who are sympathetic to psychological understanding, is that it seems to lead us inwards to the unconscious mind, into places where historians cannot follow the path of evidence. History is addressed to social life primarily, to the commonly shared meanings and the commonly occupied institutions of life; the large-scale developments, the substantial and enduring alterations which earn the name of historical change turn the historian's attention to human beings multiplied and organized. Historical evidence seems legible because it inscribes the shared language of culture and occupies the shared spaces of social life. In its simplest formulation, psychoanalysis follows a metaphoric route inwards from the literal surface to psychic space, and history is traditionally turned outwards. But psychoanalysis is not the study of silent and ineffable emotions; its 'documents' are plainly available as the typical constructions of human minds, in visual and verbal form, in action and speech, and all acts of analysis are directed to these documents of the mind.

Psychoanalysis is, in its essential interests and procedures, a theory addressed to the symbolizing activity of the mind. As one of the most influential modern interpreters of psychoanalytic theory, Marshall Edelson, puts it: 'Between stimulus and response, between event and behavior, falls the act of the mind. It is the creation of the symbol, the "poem of the act of the mind" that is the object of study in psychoanalysis.'[10] This crucial formulation, so inviting to humanist studies, offers ample space for history to operate, for the social and cultural, particularily of time and place, to enter psychological understanding. Psychoanalysis uniquely occupies the 'boundary region' touching nearly all the disciplines of the humanities and social sciences: 'the contribution of psychoanalysis to other disciplines is a conception of the mind that is adequate to the mind's complexity and to its surprising, unique, species-specific achievements in the realm of symbolization'.[11] The 'event' or 'stimulus' is the link to historical specificity; the 'symbol' may be a dream, a poem, a stylized bit of behaviour. The human mind, relentlessly and restlessly moves beyond the literal surfaces of interior and exterior reality to forms of expression that are multiform, oblique, far from obvious in meaning. The layered complexity of the mind and its innate symbolizing capacity, effortlessly deploying language and images in non-literal, tropological forms, are what account for the fullness of meaning in art, ritual, texts of every kind – and everyday 'poems' dreamt or enacted by ordinary people as well. Tracing the course of human meaning from the manifest to the latent level in systematic ways, using a stable and rigorous interpretive language, is what psychoanalysis primarily does. The forms of expression that can bear the full freight of meaningfulness are themselves endlessly various, with written texts, paintings and ritual behaviour being only some of the most obvious symbolized artefacts: dreams, conscious fantasy, colloquial speech, jokes, ritualized and improvised behaviours all carry meanings beyond the manifest. This ability of the mind to double and redouble layers of meaning into its expressive forms is the core object of pychoanalytic analysis; the processes and structures of symbolization are what this hermeneutic method analyses.

Everyone dreams in metaphor

The core theory of psychoanalysis, following Edelson's rigorous and lucid exposition, is not primarily concerned with mental health or pathology (that is its clinical application), but with the processes by which all minds produce symbolized representations to mediate between deeply felt wishes and an unaccommodating real world much more powerful than the individual. The tension between the conscious 'reality principle', which appreciates external necessity, and interior pressures of (unconscious) desire and resistance issue in a vast range of personal behaviours (including but nowhere limited to dreams) and cultural achievements (such as literature), and these products, as historians know, are legible to us because we do understand how they are made and why they have more than literal meaning. Psychoanalysis considers the mind in functional terms of ego (the conscious self), id (unconscious locus of desires), superego (carrier of cultural norms and morality) and ego-ideal (the interior measure of what we aspire to be as persons). These terms do not designate organic parts or spatial locations in the mind, but functional subsystems, each of which 'regulates the construction of and choice among symbolizations according to its own principles' (i.e. pleasure versus reality) and standards of value, such as personal gratification, mastery, the reassuring sense of social and moral conformity or self-esteem.[12]

> The id, in the interest of gratification, creates symbolic forms – 'hallucinatory' images of the gratification of sensual wishes – using transformational operations such as displacement, condensation, translation from verbal

symbols into imagery, and iconic symbolication. The 'deep,' underlying, or unconscious mental representations in psychoanalysis are often con-stituents of wishes.

The ego, in order to achieve a cognitive map of – and so master – the object world, creates verbal symbolic forms to represent that object world, using the transformational rules of natural language.[13]

The superego generates images of moral demands and condemnations. It is the internalized repository of morality, tradition, religious requirements, sexual discip-line, acquired from parents, teachers and ambient social pressures – superego lives as the ever present voice of conscience, guilt, obligation, and goads the self to both conformity and achievement. The ego-ideal foregrounds features which inspire and exemplify the ideal self formed by each individual as his or her standard for self-judgement.

The daily, nightly negotiations among the mind's systems announce themselves in bits of behaviour, the poetics of ordinary speech, and with special freedom in fantasy and dreamlife: 'dreams are a form of thought in which the representations of thought are generated by definite, distinctive canons of symbolization'.[14] Here we can briefly consider an account of a dream construction from late eleventh-century France; the 'she' of the episode is the mother of the writer, Guibert of Nogent, her son and confidant.

In the dead of a dark night, as she lay awake in her bed filled with this unbearable anxiety, the Devil, whose custom it is to attack those who are weakened by grief, the Adversary himself, appeared all of a sudden and lay upon her, crushing her with his tremendous weight until she was almost dead. The pressure began to suffocate her. Unable to speak but free of mind, she could only implore the help of God.[15]

Help came: a good spirit, perched all the while on her bed, cried out until the demon raised himself from her body and the two spirits fought until the noise awoke the maidservants. The demon fled and the good spirit paused before he left; he 'turned to her whom he had rescued and said, "Take care to be a good woman".' Guibert tells us that she treasured that admonition, 'if with God's help the opportunity should occur later'. The 'opportunity' for this special goodness turned out to be the death of her husband, when she resolved never to remarry and remained a chaste widow until her death.

This particular experience of demonic attack, recognizable in medieval terms as rape by an 'incubus' (a demon which sexually assaults women), occured to Guibert's mother, a young aristocratic wife and mother, in her marriage bed while she was left alone in charge of her elaborate household during her husband's captivity as a prisoner of war in the hands of a sadistic enemy, much feared because he often refused to ransom his captives, preferring to keep and kill them. These dire and degrading events afflicted both spouses of a troubled marriage. Both bride and husband had been married young and the marriage was a source of unhappiness and humiliation for its first several years: the young husband compensated for sexual failure in the marriage with open adultery; the wife was treated with contempt and rejection by her husband's kin and importuned with unwanted sexual advances from other men. But as Guibert repeatedly stresses, his mother was deeply religious, acutely sensitive to sinfulness and exceptionally strong-willed in persevering through familial and social conflicts. In contrast to her marriage, her long years as a wealthy independent widow were full of gratification and success. This episode, a powerful hallucinatory waking-dream, marked by paralysis of body and speech, came in the middle years of her marriage.

In its historically specific terms as a culture-bound episode combining feudal knight service, ransoms, demonic attack, an angelic spirit and celibacy as a spiritual ideal, Guibert's mother's dream is a paradigmatic trope of 'medievalness'. The socio-political framework is feudal Europe; the cultural interpretive frame is the popular belief of medieval Christianity. In keeping with gender norms of a patriarchal society, the woman is submissive and helpless, especially vulnerable in the absence of her husband, the 'natural' head of household, who should have shared the marriage bed, possessing and protecting her. Thus, a pious woman of eleventh-century France, grief-stricken over the presumptive loss of her husband, draws on her cultural repertoire to express extreme sorrow via a supernatural visitation of demonic rape from which she is rescued by a good spirit who gives her moral advice, reinforcing her religious belief and determination to conform to its values: 'be a good woman'. That, in a sketchy fashion, is a cultural paraphrase, accepting a medieval framework that combines demons, spirits and supernatural experience with patriarchy and feudal custom. But paraphrase, in the terms offered by the pervasive culture, is not interpretation: in Jonathan Lear's words, 'as soon as one approaches a dream as something that requires interpretation – that is, as something whose meaning is not immediately transparent but which nevertheless has a meaning – one needs to account both for the opacity and for the meaning.'[16] This dream, much repeated from (widowed) mother to son, is not as transparent as a cursory paraphrase suggests.

How does psychoanalysis direct us through the opacity towards this layer of meaning? It prompts us to notice the gaps and conflicts in this dreamwork fiction, starting with the interestingly cruel appropriateness of the woman's situation (husband absent and unlikely to return to the bed where she lies alone) and the experience she suffered (rape in her marriage bed): this seems like a punishment. But for what transgression? And why, after the cruel assault she had passively endured, should the departing good spirit turn on her with ambiguous warnings to be 'a *good woman*', unless her mysterious unstated sin had been terrible indeed. Pyschoanalysis also keeps us aware that the sole creator and director of dreams is the dreamer. An account of this dream must restore the dreamer as creator/subject in charge of active verbs (actions). The imperative voice is the characteristic linguistic mood of the superego: 'Take care to be a good woman'! What is the worst this woman could have done, left alone, dutifully running her marital household in chaste independence, to provoke this horror? Unless she had been enjoying herself as head of her household, secretly welcomed her imminent freedom, rejoiced at her husband's death and then immediately denied and disavowed this murderous wish (repellent to ego-ideal), disguised it as grief (a classic defence) and let the conflict enact itself symbolized in a punitive rape in her marriage bed. This information comes to us as Guibert's narrative of his mother's confidences, but he (and quite likely she) is completely silent on her husband's ransom and her reactions when he returned home. We hear only of grief and the horrific experience, no relief or joy.

The narrator skips far forwards over real events and chronological time to the *psychically* logical conclusion when, years later, the husband's actual death allowed her to fulfil the good spirit's admonition – be a 'good woman' – by remaining unmarried and sexually continent. She also becomes free of male oversight, independent, self-willed, powerful and admired. The sin implied is her secret murderous wish to be rid of her husband: the punishment was the demonic rape. The highly developed superego of this sensitive woman, perched on her bed, watching, 'sees' her innermost wishes, but at last releases her from the demon and her hallucinatory paralysis. In retrospect, as she confided this important memory story to her son, she could feel that since her husband did not die in captivity, she was not the murderous 'bad woman' of her sinful wishes. Penitent and absolved, she was free to benefit by his later death (the 'opportunity' long awaited) to become the 'good woman' of her own ego-ideal and deepest gratification: chaste, independent, spiritually elevated, head of her household. A great deal more information in Guibert's narrative memoir enriches his portrait of himself and his mother, whose personal beauty and adamant chastity he never ceases to praise and admire through the years of her strongly defended widowhood. Understanding the dream as a symbolic construct negotiated between the unconscious and conscious systems of the mind, conflating forbidden wishes, guilt, self-punishment and permission for fulfilling life desires, does not erase or entirely replace the self-interpretation of its medieval language and structure. One deepens the other, grounding cultural specificity in deep human wishes. Guibert's mother becomes a full and complex person, not a medieval ideogram.

Dreamers dream in the verbal and visual language made available by their own world, but in deep patterns shared by human beings across time and culture, which make dreams susceptible to consistent techniques of interpretation, closely related to literary interpretation. Historical specificity is contributed by the shared base for symbolization in the cultural repertoire. As Hayden White, theorist of historical writing, notes about the processes of the dreamwork and the shared domain of culture: 'symbolism, Freud says, "is not peculiar to dreams, but is characteristic of unconscious ideation, in particular among the people, it is to be found in folklore, and popular myths, legends, linguistic idioms, proverbial wisdom and current jokes, to a more complete extent than in dreams".'[17]

This may seem like introducing the theory best known for its interest in sex and neuroses from an unusually distant starting point – with demons and medieval marriage beds – but in fact it is a move that leaps directly to the centre. The special competence of psychoanalysis, developed from its classic formulations in case histories of hysteria and the texts of dreams to the work of contemporary analysts, is to describe the dynamics and structures of symbolized meaning. Whether the immediate object of these hermeneutic procedures is a person whose life is caught in repetitions of neurotic acting out, or a literary fiction, or a non-fictional memoir, the analytic instruments are the same, and the mental processes are also structurally similar whether the outcome is a pathology or a poem. Psychoanalysis is not primarily *about* mental pathology, but about the mind in its specialized capacity for symbolized expression, different for each individual, the language of the self. The interpretive procedures and trope structures of biblical exegesis, allegory, poetry of every era, are the dreamwork of Freud's classic text in their origins. The complex relations between the unconscious mind and consciousness, the restless negotiations between this deeply stratified self and the real world it cannot control or easily modify, make all forms of expression potentially complex and polyvalent in meaning. This most salient fact about psychoanalysis, that its proper object is the structure of complex meaning through symbolization, and only secondarily and accidentally concerned with mental disorders, is what brings it most readily to medieval evidence, but feels, at first, most unhistorical – until we let ourselves ask a variant of Freud's most controversial question: 'What did medieval people want?' and then let ourselves recognize their wishes, contradictions, fantasies and compromises in the evidence we have. 'From a psychoanalytic point of view, everyone is poetic; everyone dreams in metaphor and generates symbolic meaning in the process of living. Even in their prose, people have unwittingly been speaking poetry all along.'[18]

Dinner at Aunt Alveva's

We can't always be looking for demonic visitations in medieval evidence; that is too exotic and atypical, although medieval evidence offers us a rich variety. The psychoanalytically informed historian wants a wider range of those 'thoughts and visions, moods and emotions and devotions' of the more ordinary people (dare we still call them 'normal'?) Richard Southern directed medievalists to include in history. Equipped with the core insights and concepts of psychoanalysis, the historian is sensitized to the contradictions and gaps in the record of manifest life where the interior tensions 'show through'. He or she is not attempting to analyse an entire life or diagnose neuroses, but to recognize the clues which lead to a fuller sense of individuality in any record of life of ordinary people, especially caught up in conflict and cross-purposes: 'the fundamental human phenomenon to which all of psychoanalysis is a response: the fact of motivated irrationality'.[19] Jonathan Lear's term, 'motivated irrationality', behaviours and responses aimed at secret, unacknowledged or self-contradicting goals in response to great anxiety, takes us to the important concept of the defences (including denial, reversal, intellectualization, regression, repression) – the mind's creative repertoire of defence against confronting overwhelming tension. One rich and readily available narrative of 'motivated irrationality' is *The Life of Christina of Markyate*.[20] This semi-hagiography, semi-biography invites and rewards many varieties of analysis, and here I can only sketch its invitation to psychoanalytically informed reading.

The now fairly well-known twelfth-century text tells the story of an upper-class girl from a small English town who resisted her family's strenuous attempts to get her married, ran away to become a religious recluse, and was in due course the subject of a Latin *vita*. She was born some time in the last years of the eleventh century and lived until at least 1155 and perhaps a decade further. The town of Huntingdon was her home and the nearby male monastery of St Albans was the scene of her first religious fervour, as well as giving her mentors and friends throughout her life and, at last, a biographer who knew her personally and was one of her staunchest supporters. Women like Christina needed supporters because they always had detractors. Celibate female spirituality surrounded by celibate male guidance, instruction, support and, in the end, writing, is a distinct cultural topos of the age. In Christina's case it is particularly well marked because custody of her body and soul, and notably her reputation, was disputed between carnal and continent males, and between hostile gossip and reverent report.

The current pedagogic rubric for reading this text is *gender*: gender as an analytic instrument for unpacking the layers of predetermined expectations and restrictions, which, in traditional patriarchal societies, closed around the biological female, virtually at birth, to force her into conformity with a social template of lifelong obligations. At about the age of 13 or 14, Christina made a religious commitment of her body in the form of a secret vow of perpetual virginity during an inspiring visit to a monastery, which would seem an aspiration that matched the universal ideals for feminine spirituality in her society, to be greeted with applause and support. But when this vow was revealed to her parents a few years later, it exposed the vicarious ambitions her family had invested in her and the powerful cross-current of emotions focused on her, and caused her family to turn floridly disfunctional, drawing into the domestic psychodrama numerous others attracted to the 'Christina problem'. And in the end she won, which makes hers a rare female success story. Her *vita*, an openly partisan narrative by a monk of St Albans, approaches us already heavily interpreted in its emplotment and diction (following a generalized hagiographic model): self-dedicated young virgin resists evil worldly seducers and escapes intact to spiritual freedom. This text is filled with conflict: sexual desire, sublimation and resistance, struggle between the sexes, parent–child love and rebellion, erotic drama on many levels.[21] Gender theory alone is not equal to the task of unpacking the depths of this narrative. Depth psychology registers the personalities who parade through the text, often badly out of step with ego-ideal and social norms. The narrative brings together a cast of characters with evident, vociferously forwarded aims – and less manifest wishes, raised to visibility through the cross-purpose gaps and ellisions of the text, which is itself the rhetorical or manifest level of the action swirling around the node of conflict that is Christina. A striking feature of Christina's life is that no one connected with her story seems able to have what they ostensibly demand, or to acknowledge what else they might want.

The central characters include Christina; her mother Beatrix; her father Autti; Ranulf Flambard, the current bishop of Durham and former lover of Christina's Aunt Alveva (sister to her mother); Burthred, a wealthy young man hoping to be married to Christina; and numerous others with strong Christina-related feelings. This richly detailed text cannot be fully conveyed here, but one event touched off many repercussions:

> While Ranulf the bishop of Durham was justiciar of the whole of England, . . . but before he became a bishop [1099], he had taken to himself Christina's maternal aunt, named Alveva, and had children by her. Afterwards he gave her in marriage to one of the citizens of Huntingdon and for her sake held the rest of her kin in high esteem. On the way from Northumbria to London and on his return from there he always stayed at her house. On one occasion when he was there Autti, his friend [Christina's father], had come as usual with his children to see him. The bishop gazed intently at his beautiful daughter, and immediately Satan put it into his heart to desire her . . . he had the unsuspecting girl brought into his chamber where he slept . . . Her father and mother and the others with whom she had come were in the hall giving themselves up to drunkenness.[22]

Alveva, the hostess, never appears in the action, which all takes place in the bishop's bedroom, where Ranulph propositioned Christina, who tricked him and managed to lock him in the bedroom, thus setting off the complicated train of misadventures, the familial war of the wills, that culminated in her religious career.[23] The expanded social/cultural setting for this vignette opens out to encompass conflicting systems of secular patriarchal lineage – based on ambition, with marriages as key strategic moves (and virgin daughters as pawns) – against the doctrinal/institutional claims of the post-reform Church to enact its spiritual superiority in the world (with virgins as occasional icons).

In the midst of these monolithic systems of control, focused with such intensity on the female mind and body, is Alveva, who had openly dishonoured her family, religion and personal honour throughout her immoral youth, with illegitimate children (and maybe some legitimate ones together in the same house), living in the same small town as her respectable married sister, in a luxurious house (the narrator notes the nice tapestries in the guest room), with a rich husband, furnished by her ex-lover, who visits them quite regularly and openly. And what is this fugitive from the hegemonic systems of moral discipline doing? She is giving a dinner party, with everyone there, as usual: husband, ex-lover, sister, brother-in-law, unmarried nieces and others. The bishop liked all her relations, 'for her sake', we are told with breathtaking casualness – a shred of circumstantial explanation slipped in as the narrator hurries on to more interesting matters. The *vita*, and the scholarship on it, concentrates on Christina, but her Aunt Alveva speaks to us provocatively of a life lived 'against the grain'. Alveva only re-enters the story after her three sentences as the sympathetic unnamed 'aunt' who actually helps the unhappy Christina run away from home. Alveva is not a heroine, but a little forcefield of personal ambitions, audacities, pleasures sought and frankly enjoyed within a social system deployed mainly to contain and discipline her. Without her assistance at a crisis moment, Christina could not have escaped to be heroine of a Latin *vita*. The strange aura of tolerances and withheld judgements, for all the indubitable gossip (a bishop's entourage hardly enters a small town secretly) that opens out around Alveva, exposes the fact that desires and ambitions like hers, her compromises and contrivances, using and being used by men, culminated in her sympathy with a rebellious niece who didn't want to be handed from father to husband. She was the only family member who supported a girl's defiance.

Alveva, once one pays attention to her, in her startling individuality, as a nexus of everything stereotypically not medieval, encourages us to reread Christina and her entire situation and notice the developed and repeated pattern of Christina's intense and important erotic relations with men, which invites a psychological analysis. The approach so near to sexual danger, close physical proximity with men who frankly wanted her, the ultimate defence of her chaste body achieved in a dense atmosphere of desire and threat, was Christina's signature pattern, often repeated.[24] Her biographer offers us these episodes as evidence of her reiterated triumphs over temptation and threat; his interpretation is inscribed in his diction, narrative patterns, assumptions. It is addressed to conscious intention and cultural ideals. Only if consciousness is all there is, would a twelfth-century religious interpretation be wholly sufficient.

This is not a search for more medieval neurosis. What I want to consider is the complexity of normalcy, everyday normalcy, as exhibited by everyone in this text and how we think about them behaving as part of the society of a small English town early in the twelfth century. It is an interesting time: well into the orderly, peaceful reign of Henry I; the new strict standards of the papal reform movement becoming known and accepted; enhanced intellectual and religious refinements connected to general prosperity; little in the way of heresy or widespread dissatisfaction with traditional orthodoxy and morality. In general, we have every right to ascribe to the Huntingdon patriciate, Christina's social circle, the ordinary range of beliefs and attitudes, moral constraints and social disciplines considered quintessentially 'medieval': emphatically serious about sin, salvation and social propriety. These people were not low enough to evade knowledge of their obligations to God and society; they were not high enough to transgress in the cavalier way of wicked royalty. Not one of them whose actions attracted the author's interest conformed entirely to a moral or social template.

Beatrix alternately berates and bribes her daughter, demonstrating the alluring pleasures of a rich woman's married life, like her own, and then viciously urging her betrothed to rape her when Christina rejects her mother as a model. Autti is furiously hurt as his favourite daughter defies his natural authority, and openly expresses his humiliation and sense of rejection, and yet Christina is never without money and is entrusted with the household keys – a signal honour. The father–daughter relation in this text is fraught with powerful emotions and is replicated in Christina's relations with other senior and authoritative males who wish to love and control her. And without going deeply into the fascinating complications of Ranulf Flambard's life (he got a bad press from the monastic chroniclers), suffice it to say that he was mad, bad and dangerous to know for nearly everyone but his many friends who loved him, and he took care of all his children, of whatever provenance. His reactions, even filtered through this moralistic set-piece of the foiled evil seducer, are curious.[25] A 17-year-old girl made quite a fool of him and he reacted by bringing her expensive presents (there is an apology in that gesture, even if he still wanted to seduce her). And when she still rejected him, what the author terms his 'revenge' took the form of finding her a husband who was young, rich and noble and starting to build a new house for her there in Huntingdon, near her family. The biographer's reading of this is that if the bishop couldn't have her, he wanted to take her virginity by proxy, as it were, which is a very interesting interpretation in itself from a twelfth-century monk. But it does not require any strenuous wrenching of the imagination to look at these actions and see something strangely mild, some mix of incommensurate aims, from this notorious arrogant sinner who was on such jolly terms with his ex-mistress and everyone connected to her. Far from a proxy rapist, the rich husband was more likely a munificent apology directed to Autti and Beatrix, and to Alveva, whose home he had insulted.

Altogether in their simultaneous mismatched and contradictory intentions (swearing perpetual virginity; staying so late in the bishop's bedroom; plotting seductions; arranging marriages and dinner parties; rejecting silk gowns), all the personages of Christina's world pose yet again our Freudian question: What did these medieval people *want*? And how do we think of them in their entirety as beings who *want*, as human beings with beliefs, desires, conscious purposes and intentions, many not wholly available, or acceptable, to consciousness?

Do we need an analysis of the psyche to investigate the everyday world of the normal, which is what this family drama of clashing authorities and ambitions is? Every one of our patently normal people vacillates and contradicts himself and herself, reveals the missteps of complex self-awareness and unawareness: Bishop Ranulf roughly grabs Christina, then offers her silk and jewels, then a desirable husband; her parents enforce their traditional unquestioned authority by alternating yelling and threats with pleading and promises, lavish presents and banquets and parties, where Christina's mother once beats her severely. Christina austerely vows a virgin life, but somehow gets betrothed and even says the marriage words before she changes her mind; her husband-in-name creeps into her bedroom with the express intention of marital rape and then sits docile and quiet on the bed with her as she tells him stories of St Cecilia and other holy virgins.

One of the less acknowledged powers of psychoanalytic theory is that it distinguishes so interestingly between normalcy and conformity. By including the unconscious mind in its persistent pressure on consciousness, psychoanalysis offers a thick description of the 'normal' which is not the same as normative or submissive; the psychoanalytically normal mind is not an ideogram of hegemonic cultural interests. One analyst, with far-ranging experience of cultural variation, remarks on 'the commonly held illusion that the neurotic, the psychotic, and the delinquent are hard to understand. Actually, such individuals are much simpler, much more dedifferentiated, and considerably more derivative than normal persons are; they are much less colorful, imaginative, original, and individualized.'[26] The mind, stubbornly harbouring its own projects, can 'push back', even when the world is coherently and omnipresently organized to guide it to conformity. The central issue for historians is that only a theory of mind which reaches beyond the manifest, culturally coded consciousness is *interpretive*. Social constructionist theory, for example, can offer only cultural paraphrase. Interpretation requires a language which addresses itself to deep resources of language, including the non-verbal but rule-governed languages of gesture, behaviour and exchange. Without a depth psychology we are ill-prepared to invite ourselves to Alveva's dinner party and appreciate the complicated banquet of feelings, beliefs, desires and memories that was served there.

> One way of describing Freud's conception of the mind is to say this it is based on the primacy of the will, and that the organization of the internal life is in the form, often fantastically parodic, of a criminal process in which the mind is at once the criminal, the victim, the police, the judge, and the executioner.

This is Lionel Trilling invoking Freud in the preface to Dicken's *Little Dorrit*, a place both surprising and quite right.[27] He speaks in metaphor, as is appropriate to a theory primarily concerned with symbols and the mental processes of symbolizing. Trilling was among the many, albeit not enough, students of psychoanalysis who unerringly located the core of the theory in the mind's capacity to create symbolic worlds, and thus understood the intimate relation between psychoanalysis and literature:

> For, of all mental systems, the Freudian psychology is the one which makes poetry indigenous to the very constitution of the mind . . . [It] was left to Freud to discover how, in a scientific age, we still feel and think in figurative formations, and to create, what psychoanalysis is, a science of tropes, of metaphor and its variants, synecdoche and metonymy.[28]

This is the essential point, especially for cross-disciplinary studies: 'a science of tropes'. A science of tropes is culturally conservative in the best sense: it preserves and recapitulates the enduring insights of classical rhetoric, of scriptural and literary exegesis, and brings them forward to the modern discovery of their central role in mental life. Literature, the most complex kind of symbolic construction, is intelligible to us across centuries of distance and difference because it proceeds from the same human resource of mental processes, however disguised and recoded by culture. Scholars regularly find that human artefacts of language, such as *The Canterbury Tales*, are saturated with polyvalent meanings, self-reflexive ironies, complexly related strata of meaning, compressed and shadowed significations, endless ways of conveying more than literal meaning. These deep and complex levels of meaning in language are understood as 'really there' – the expression of sensitive and complex minds from centuries ago, not merely the clever invention of modern readers. By acknowledging the complexity of literature from the ancient and medieval past, we are already implicitly committed to a concept of the mind which supports such expressive depths, and to a theory of mental life in which symbolizing is a primary activity.

The core theory

Psychoanalysis permeates contemporary culture; everyone already has some understanding of it, and a course of well-chosen readings can marshal and organize this near-knowledge to good effect. What the historian generally aims at is not total or intensive analysis of a life or text, but an approach to evidence sensitized to the 'motivated irrationalities' always present in life and literature. For medievalists, there is extraordinary interest in approaching psychoanalysis from classical philosophy onwards, rather than the usual move from modernity back to premodern society; Jonathan Lear's studies of ancient thought and the *longue durée* human project of understanding our own minds, with Freud integrated clearly in a tradition that begins with Plato and Aristotle, are a good place to start. *Love and Its Place in Nature: A Philosophical Interpretation of Freudian Psychoanalysis* (1990) and *Open Minded: Working Out the Logic of the Soul* (1998) are indispensable. Historians will find his discussions, 'The Interpretation of Dreams' and 'What is Sex?' (both in *Love and Its Place*) lucid and good to think with, and his concept of the 'Idiopolis' – 'that even when a person participates in shared cultural activities, those activities will tend also to have an idiosyncratic, unconscious meaning for that person' – of striking pertinence for medieval culture.[29] Interestingly, Lear's immersion in classical culture makes psychoanalysis appear a familiar and inevitable development of human thought:

> Psychoanalysis is an extension of our ordinary psychological ways of interpreting people in terms of their beliefs, desires, hopes, and fears. The extension is important because psychoanalysis attributes to people other forms of motivation – in particular wish and phantasy – which attempt to account for outbreaks of irrationality and other puzzling human behavior.[30]

It should be clear that psychoanalysis is not limited to Freud's work, but all post-Freudian developments are firmly grounded in the classic theory. Marshall Edelson's years of clinical and theoretical work, rigorous, logical and sensitive to applications far outside analytic circles, have produced writings of unusual depth and clarity, exactly the guidance needed by historians wanting sufficient and well-grounded knowledge. He locates the 'core theory' in the following works by Freud: *The Interpretation of Dreams, The Psychopathology of Everyday Life, The Three Essays on the Theory of Sexuality* and *Jokes and their Relation to the Unconcious.*[31] These works contain the central concepts of the foundation of psychoanalysis, but they are not read as canonical or fixed; Edelson's own refinements on the core theory, acknowledging the work of many other thinkers in the field, are recommended for their rigour and clarity. These include the invaluable, already cited *Psychoanalysis: A Theory in Crisis*, as well as his *Language and*

Interpretation in Psychoanalysis (1975) and *Hypothesis and Evidence in Psychoanalysis* (1984). The work of psychoanalysis for medieval history, generically put, is to restore our sense of human recalcitrance, to locate the source of the 'pushing back' against a conformity-demanding culture and to respect the self, albeit medieval, by acknowledging 'the psyche's fundamental activity: to inform the world with meaning'.[32]

Guide to further reading

The number of books, articles and specialized journals available on psychoanalysis is daunting enough to discourage an interested beginner. However, the historian wishing to acquire a cogent grasp of the central concepts of psychoanalysis and how to deploy these as techniques for reading and interpreting historical evidence can do very well under the guidance of a few clear and reliable writers, and proceed by following their recommendations in bibliographies and footnotes. It is important to remember that interesting post-Freudian developments like object relations theory and the work of Jacques Lacan cannot be understood with any clarity without a grounding in the basic concepts of Freudian theory.

I have found the most rigorous and lucid guide to psychoanalytic theory in the work of the eminent analyst/theorist, Marshall Edelson. He not only offers the clearest explanation of exactly what it is that psychoanalytic theory examines and explains, but his interests in literature and music also make him a valuable guide to using psychoanalysis for work in other areas. His explanation of the meaning and role of sexuality in psychoanalytic interpretation is exceptionally lucid. Among his several books and many articles, the most comprehensive is *Psychoanalysis: A Theory in Crisis* (Chicago, IL, and London, 1988); for those interested in the intellectual structure of psychoanalytic thought, see *Hypothesis and Evidence in Psychoanalysis* (Chicago, IL, and London, 1984).

Edelson focuses helpfully on 'the core theory' of Freudian thought and directs readers to begin with the following works by Sigmund Freud (all are easily available in many editions): *The Interpretation of Dreams*; *The Psychopathology of Everyday Life*; *The Three Essays on the Theory of Sexuality*; and *Jokes and their Relation to the Unconscious*.

Freud himself offered some of the clearest short summaries of his work. See *Introductory Lectures on Psychoanalysis*; *An Outline of Psychoanalysis*; and *Five Lectures on Psychoanalysis*.

The place of psychoanalysis in western culture, what permanent aspects of human life and inquiry it adresses, and why it is both essential and yet endlessly contested, are issues discussed with a very appealing combination of erudition and wit by Jonathan Lear, a scholar of classical philosophy and psychoanalysis. Working out the *longue durée* connections between ancient thought and psychoanalysis, Lear speaks with special intelligibility to medieval historians. See *Love and Its Place in Nature: A Philosophical Interpretation of Freudian Psychoanalysis* (New Haven, CT, and London, 1990) and *Open Minded: Working Out the Logic of the Soul* (Cambridge, MA, 1998).

For historical work, we need techniques for bridging the space between psychoanalytic theory and historical evidence – ways of turning one kind of explanatory narrative into another. Roy Schafer, an analyst with great interest in the role of linguistic form in psychic life, is exceptionally helpful. Among many publications, see *A New Language for Psychoanalysis* (New Haven, CT, 1978); 'Narration in the Psychoanalytic Dialogue', *Critical Inquiry* 7 (1980–1), pp. 29–53; and 'The Rhetoric of Displacement and Condensation', *Pre/Text* 3 (1982), pp. 9–29. On rhetorical tropes and psychoanalysis, see Hayden White's essay, 'Freud's Tropology of Dreaming', in his *Figural Realism* (Baltimore, MD, and London, 1999), pp. 101–25.

I have discussed psychoanalysis and medieval history in 'No Sex, No Gender', in *Studying Medieval Women: Sex, Gender, Feminism* (Cambridge, MA, 1993), pp. 117–41, with an appendix of readings; and have applied psychoanalytic concepts to the records of medieval lives in 'Reading the Book of Margery Kempe', *Exemplaria* 3 (1991), pp. 29–66, and 'The Family Romance of Guibert of Nogent: His Story/Her Story', in *Medieval Mothering*, edited by B. Wheeler and J. Parsons (New York, 1996), pp. 359–79.

Notes

1   Jacob Burckhardt, *The Civilization of the Renaissance in Italy*, introduction by Benjamin Nelson and Charles Trinkaus (New York, 1958), p. 143.
2   Nelson and Trinkaus, introduction, p. 3 and n. 18.
3   Lee Patterson, 'On the Margin: Postmodernism, Ironic History, and Medieval Studies', in the New Philology issue of *Speculum* 65 (1990), pp. 87–107, especially his scathing comments on the complacent appropriation of interiorized subjectivity for post-medieval persons by scholars whose own projects depend on defining 'modernity' as a quality of their specialist fields, pp. 95–100. I discuss some of these same issues in 'Did Mystics Have Sex?', in *Desire and Denial: Sex and*

Sexuality in the Premodern West, eds Jacqueline Murray and Konrad Eisenbichler (Toronto, 1996), pp. 296–311.

4   Eric Jager, 'The Book of the Heart: Reading and Writing the Medieval Subject', *Speculum* 71, pp. 1–3, for the pre-modern metaphors of interiority.

5   Mary Carruthers, *The Craft of Thought: Meditation, Rhetoric, and the Making of Images, 400–1200* (Cambridge, 1998).

6   Mark Solms, 'Freud Returns', *Scientific American* (May 2004), pp. 82–8, and more directly from the scientists, see Drew Westen and Glen O. Gabbard, 'Developments in Cognitive Neuroscience: 1. Conflict, Compromise, and Connectionism', *Journal of the American Psychoanalytic Association* (2001), pp. 50–98, on latest developments in integrating cognitive neuroscience with psychoanalytic theory, with extensive bibliography. Of particular interest to students of historical evidence is the work of Matthew Hugh Erdelyi, *The Recovery of Unconscious Memories: Hypermnesia and Reminiscence* (Chicago, IL, 1996), which combines clinical psychology (experiments) and psychoanalysis to unpack the processes by which memory is lost, constructed and recovered: 'Interestingly, the [current] controversy recapitulates on a vast scale a crisis confronted almost a century ago by Sigmund Freud in the solitary confines of his own clinical practice. And the emerging consensus, led by experimental psychology, confirms Freud's much maligned (and often misrepresented) conclusions: unconscious memories can be recovered; unfortunately, some the recollections are false, even if some of them are true, and even recollections tending to be true are typically garbled and overlaid by distortions (pp. xiii–xiv).

7   Jonathan Lear, *Open Minded: Working Out the Logic of the Soul* (Cambridge, MA, 1998), p. 18.

8   Lear, *Open Minded*, p. 18.

9   Richard Southern, 'The Shape and Substance of Academic History', in Richard Southern and Robert Bartlett, *History and Historians: Selected Papers of R.W. Southern* (Oxford, 2004), p. 100. This drily comic and deeply humane essay by the Chichele Professor should be required reading by every historian, for his account of how history struggled into existence as a discipline and clung to constitutional history for its only claim to rigour and coherence.

10 Marshall Edelson, *Psychoanalysis: A Theory in Crisis* (Chicago, IL, 1988), p. 20, where he refers to the poet Wallace Stevens for the phrase, 'poem of the act of the mind'.

11 Edelson, *Psychoanalysis*, p. 62.

12 Edelson, *Psychoanalysis*, pp. 13–15 for the systems-subsystems model of the psychoanalytic account of the mind.

13 Edelson, *Psychoanalysis*, p. 16.

14 Edelson, *Psychoanalysis*, p. 47.

15 The most recent translation is that of Paul J. Archambault, *A Monk's Confession: The Memoirs of Guibert of Nogent* (University Park, PA, 1996), pp. 40–41.
16 Lear, *Open Minded*, p. 85: 'And once we recognize that a mind has to be capable of making (what from the perspective of secondary process appear to be) strange leaps and associations, we see that a mind has to have something like displacement and condensation as forms of mental activity. For displacement is the bare making of associations by linking ideas; condensation is the bare making of associations by superimposing them. These activities both discover and create similarities'.
17 Hayden White, 'Freud's Tropology of Dreaming', in *Figural Realism* (Baltimore, MD, 1999), p. 122.
18 Lear, *Open Minded*, p. 31.
19 Lear, *Open Minded*, p. 54.
20 *The Life of Christina of Markyate*, trans. and ed. C.H. Talbot (Oxford, 1959). The anonymous author of this semi-hagiographical biography was a monk of St Albans during the life of Abbot Geoffrey, Christina's close friend and financial supporter; the monk was acquainted with his subject and her family at least during the later years of her life in the mid-twelfth century.
21 I have analysed the narrative gaps, self-contradictions and wilful elisions in 'Christina of Markyate and Theodora of Huntingdon: Narrative Careers', in *Reading the Middle Ages: Essays in Honor of Robert Hanning*, ed. Robert Stein (University of Notre Dame, forthcoming); and a groundbreaking article which examines the evidence that can explain the near total lack of a cult for this ostensible saint is Rachel M. Koopman's 'The Conclusion of Christina of Markyate's Vita', *Journal of Ecclesiastical History* 51 (2000), pp. 663–98.
22 *Life of Christina*, pp. 41–3.
23 To fill in the plot for those unfamiliar with the text: Bishop Ranulph brings Christina expensive presents on his return trip from London, either as another attempt at seduction, as the author suggests, or by way of apology, as I infer, but she refuses to accept them. Next, the bishop encourages a wealthy young man named Burthred to offer marriage to Christina; her parents urge her to accept, but she announces that she is vowed to perpetual virginity. Family conflict ensues, with an ambiguously coerced marriage, Christina's continued rejection of Burthred and, finally, her flight from home to the custody of religious protectors.
24 See the episodes in Christina's life with the unnamed ecclesiastic with whom she fell in love (pp. 115–19) and her later ambiguously erotic friendship with the abbot of St Albans (pp. 135–75).
25 For Ranulph's truly flamboyant career, see Richard Southern, 'Ranulf Flambard', in *Medieval Humanism* (Oxford, 1970), pp. 183–205.

26 George Devereux, *Basic Problems of Ethno-Psychiatry* (Chicago, IL, 1980), p. 157.

27 Lionel Trilling, preface to Charles Dickens' *Little Dorrit* (Oxford and New York, 1966, many reprintings), pp. vii.

28 Lionel Trilling, 'Freud and Literature', in *The Liberal Imagination* (New York, 1951), pp. 52–3.

29 Lear, *Open Minded*, p. 69.

30 Lear, *Open Minded*, p. 25.

31 Edelson, *Psychoanalysis*, p. 64.

32 Lear, *Open Minded*, p. 70.

Part 2

Literary techniques for reading historical texts

4

Literary criticism and the evidence for history

Robert M. Stein

The so-called 'linguistic turn' in history happened long enough ago that apocalyptic fears about the imminent disastrous end of the profession can be assumed to have been refuted by the mere passage of time. Largely a development in England and North America in the 1980s, we can date it, at the latest, to 1987 and John Toews' influential review essay, 'Intellectual History after the Linguistic Turn: The Autonomy of Meaning and the Irreducibility of Experience',[1] whose title announced its accomplishment even as the body of the essay worried over its significance. Toews importantly points out that what the linguistic turn signified for the historian is not merely the parasitic use by one branch of history writing (intellectual history) of methods and perspectives properly belonging to another discipline (linguistics and literary criticism) or, even worse, the imperial colonizing of the discipline of history by the upstart pretender, literary analysis. The linguistic turn that history took in the 1980s was rather a local effect of a large-scale shift in emphasis that occurred throughout all the disciplines constituting the human sciences.[2]

This large-scale shift, we can summarize, was part of a redirection of academic inquiry that emphasized the question of how meanings come into being and their relation to experience. Meanings, this inquiry reveals, are the result of complex social and psychological operations that occur constantly at various levels in culture and society.[3] And the shift of attention to these operations as a matter for philosophical, anthropological and historical investigations thrusts the fact of language, and specifically of writing, into the foreground, for the production and dissemination of meaning takes place entirely within the symbolic sphere of expression and communications. In the process, literary criticism was transformed from a set of techniques specific to understanding the great artistic works of high culture, whose linguistic procedures are complex, evocative, highly self-conscious and subject to infinite interpretation, to an indispensable part of any historical investigator's toolbox. For not since the biblical Adam first gave names to all the birds and beasts has a human being lived in a non-verbal world; rather, humanity is enmeshed in the webs of significance made from a world constantly spoken about, written about and argued about in language. Description of life is not passively recorded, but constructed. Experience is thus not something that happens 'outside' of language, something that language can follow after in order to give a more or less truthful account. Rather, experience – the experience from moment to moment of an individual or the collective experience of a generation or of an age – is something that always occurs in a world already spoken about, a world already saturated with meanings, already filled with language. Language is thus, in the first instance, always implicated in experience. And yet, individual and social experience is neither a totally linguistic phenomenon nor is it reducible to what is said about it: there is always an excess, something that seems to escape any account of what has happened and what that happening means.[4] It is this excess that opens the possibility of continuous interpretation, for no word is the last word, no account, however thorough, is ever the definitive account.

For the practice of history, the focus of attention on those operations in culture and society that create meanings in the medium of language has significant implications for three primary areas of activity: for the meaning of 'evidence', for the meaning of 'context' and for the fact that what the professional historian actually does, in order to do history at all, is to write a text, be it a book, an essay, a review or a class lecture. As we examine each of these spheres, what we shall need to attend to is a continual transaction between the new utterance and the language with its embedded values and descriptions that has been there all along.

The source as document

Let us begin by being painstakingly obvious. Historians work from evidence to reconstruct, as accurately as possible, the life of the past. Working from evidence and not making assertions beyond what can be supported by evidence are what constitute historical research as a discipline. We learn very early on to divide our information into primary and secondary sources. As we all know, secondary sources consist of material written *about* the entity we want to study, and primary sources are *of* the entity in question – produced by historical contemporaries. Primary sources are our evidence, for they are bits of 'past' material – usually but not invariably written material – still extant in the present. They are the products of a fuller past reality that no longer exists (that it does not exist *now* is, of course, what makes it past and what necessitates its reconstruction if it is to be known at all), even though fragmentary pieces of it still exist in the present and may indeed have present uses, different from their uses at the time of their creation. Primary sources are thus traces of the past in the present, and historians use them to create an account of the fuller past that no longer exists and to which, as evidence, they provide a point of access. With certain exceptions – time capsules, suicide notes, certain kinds of archival records designed specifically for an anticipated posterity – historians use things for sources that were never intended by their makers to be sources. The source is the result of human activity; it is the material remains of the motives, intentions, aims, desires or plans of those who brought the 'source' into being to accomplish with it certain things in their own time: someone desired to distribute her property in a particular way after death and so made a will; someone else wanted to tell his sister in London about his first day at university and so wrote a letter home; someone planned to go to market and so wrote a shopping list. They did not intend these things to be 'sources' for us but rather to do certain things for themselves: to allay the anxieties of a loved one, for example, or to organize their own affairs.

I have described this obvious and well-known meaning of 'primary source' in what might seem to be an excessively obvious fashion precisely because it provides us with an important entry point to understanding the relevance of literary theory and criticism to the work of the medieval historian. I hope we can see from my description that nothing is a source as such; a source becomes a source only as it enters into a transaction with a historian to serve the historian's purposes, when it is used, in other words, as 'a document'. And this can be true even of those sources that are created with posterity in mind, for some of the most interesting historical documentary study occurs when historians use sources for purposes other than those for which they were intended.[5] What I cannot stress strongly enough here is the relational quality in what we conventionally call a source. Put most precisely, a historical source is not strictly an isolated entity, static or frozen in time, but exists now as a relation and in an act of reading. It is a relation between a present entity (let us get to the heart of the matter instantly and call it a text), a present reader of that text (in this case, the historian) and a disciplinary structure (in this case, history) that supplies the reader with an interpretive context, a purpose for reading and a protocol for interpretation. The source is a social fact, and one fully mediated by language.

Historians thus typically read sources as documents: pieces of the material reality of the past that more or less reliably provide a more or less clear window onto the world of the past that is the historian's object of study. And the profession has elaborated a fairly complex protocol of reading – source study – to maximize the reliability of the source for the historian. The language in which source study is described is typically judicial: the source, one says, is put into the dock, it is examined or questioned, it is made to reveal the truth, to become a trustworthy witness. And using the account that the historian is able to draw from the source, the historian, in turn, can construct a new account of what the source has revealed about the social world which produced it.

As soon as one considers the historian's activity dynamically, literary criticism enters historical work at all points. For what I have just described as source study is in fact a complex act of reading. Although evidence can be any artefact whatsoever, in a highly literate culture like the European Middle Ages, evidence for working historians is primarily composed of written artefacts – laws, wills, administrative records of all sorts; as well as memoirs, letters describing people, places and events; a great variety of written accounts, ranging from informal narratives to self-consciously written history. Some of this evidence can be analysed by sociological or quantitative methods, but most evidence must be interpreted by other, more qualitative means.

In order to be a document at all, the source is almost always first and foremost a text (for not only written materials, but also social practices, such as religious rituals or social structures like kinship groups are texts in so far as they too are themselves part of the symbolic system by which a culture constructs its own meanings for itself). Historians tend to leave purely physical evidence, such as potsherds and kitchen middens, to the professional eyes of archaeologists. And to be a text at all, it must always be a text among other texts.[6] Calling the document a text emphasizes its structure of signification, capable of being read in various ways for a range of purposes and always subject to the properties of the symbolic system of which it is an instance. A text gets itself turned into a document in a very specific situation. By being read in a particular way according to a particular protocol, an interpretive procedure sanctioned by the discipline and its traditions, the historian interprets the text as a piece of past reality that reveals more of the past than it contains. Its particular meaning for the historian (as a document) occurs at the site of reading, not of original writing: it is not how it came into being but how it is read as a text among other texts that transforms the text into a historical document.

The source as text

Let us consider a concrete example. Here is an entry from the Hyde Chronicle, a monastic annal, described as generally brief, dry and businesslike.[7] It is the sort of chronicle entry that historians like, one whose language seems to be sufficiently transparent to cause few if any literary problems. The year is 1066:

> *Pridie idus Octobris, ipso die Sabbati, factum est bellum in Anglia inter Normannos et Anglos, in quo bello quamvis varius in primis utrimque fuit eventus et nulla morientium requies, tandem manifesto Dei judicio eodem die rex Haroldus, corruens morte sua et bello et Anglorum regno finem imposuit.*

Willelmus igitur Comes, potita victoria, ipso sequenti die Natalis Domini apud Londoniam elevatus est in regem, finitumque est regnum Anglorum et inchoatum regnum Normannorum.[8]

[On the day before the Ides of October, a Saturday, war was made in England between the Normans and the English, in which war although at first the outcome was indecisive for both sides and there was no respite from dying, finally by a manifest judgement of God, on that same day King Harold, rushing to his death, brought an end both to the war and to the reign of the English. Count William, having won the victory, on Christmas day following, was elevated to the kingship in London, and the reign of the English ended and the reign of the Normans began.]

This entry seems straightforward enough. Experts have dated the composition of this section of the chronicle to the reign of Henry I, within living memory of 1066. But as a source for understanding what went on at Hastings, this chronicle entry is disappointingly spare. It is suitable perhaps only to document the date (14 October), the duration (one day) and the decisive event (the death of King Harold) that brought the battle to an end, and thus to corroborate the more elaborate (and hence the more suspect) literary accounts of such eleventh- and twelfth-century sources as William of Poitiers' *Gesta Guillelmi* or Orderic Vitalis' *Historia Ecclesiastica.*[9]

To write the previous paragraph I have in fact performed a rather radical surgery on the chronicle entry. To read this chronicle entry as merely documenting the date, the duration and the event, I consider only a very small part of the content of the entry and, employing fairly ordinary standards of source study, I skip over the rest as irrelevant for my purposes. Employing as it does the language of miracle, portent and prodigy on the one hand (*manifesto Dei judicio*) and the most schematic and stereotyped indication of battle on the other (*quamvis varius in primis … eventus et nulla morientium requies*), a good half of this chronicle entry falls outside the professional belief structure of a working historian and cannot be taken for a transparent window onto 'what happened'. That the chronicler calls the event 'a manifest judgement of God' explains nothing for us; rather it is the sort of thing that generations of historians have dismissed as medieval credulousness and superstitious belief; the sort of thing that has made medieval chronicles so notoriously unreliable as sources. And his account of the battle as indecisive at first, with much death on both sides, looks like the kind of fictionalizing elaboration endemic to texts of this sort. Was the battle really indecisive at first, we ask, and when did it become clear that the Normans had the upper hand? Reading this chronicle thus presents a general historiographical problem related to, but significantly larger than, the question of its documentary reliability, for in its claim truthfully to represent something that happened 'in reality', independent of the confines of the narrative, historical writing requires the reader's assent to its own formal techniques of representation. And as we have just seen, a significant piece of this narrative stands outside those canons of truth that gain our routine assent; we pass over a great deal of what the writer chose to say in our search for a fact on which we may rely. What are the precise grounds by which we accept a part of an ancient narrative as providing a clear window onto the past and dismiss the rest?

We can begin to address this question by making a banal observation, but one with large and complex ramifications. For any experience whatsoever to be represented in a medium such as language, the experience needs to conform to a variety of constraints specific to the medium. One might think here of the mechanical constraints imposed by the huge apparatus needed for even the simplest shot in classic cinema. Equally important in film are the less directly material constraints imposed by the conventional vocabulary of cinematic representation. So too with regard to the more abstract and flexible medium of written language and especially with regard to what seems at first glance the 'natural' and entirely unproblematic practice of representation that we have been looking at – a practice of storytelling that seems to be a simple *mirroring* of experience in a transparent medium that draws no attention to itself as either a practice or a medium. Yet in dismissing much of the chronicle entry as unreliable – either superstitious or made up – what we are actually doing is judging its plausibility *to ourselves* and not its truth. Since wars for us are likely to begin on a particular day that we can name, a Saturday for example, the chronicler naming the day has the ring of truth; to assert that we know the will of God is less likely to provoke immediate assent from us *as professional historians now*, though it might well be quite plausible for the chronicler's own audience, as well as for a large segment of the population of our own contemporary world. We need to direct our attention differently. Rather than picking and choosing only those aspects of the source's content that are plausible for us to believe and because they are plausible seem true for us,[10] let us try to look *at* this chronicle entry rather than *through* it. How does this little narrative go about the business of securing belief in its representation of experience?

First of all, the chronicler uses the language of the calendar (*Pridie idus Octobris, ipso die Sabbati ... eodem die*) to insist that the whole event took place on a single day on which war broke out (*factum est bellum*) and came to an end (*bello ... finem imposuit*). The one-day war is thus absolutely decisive. The language of the calendar is also used to connect this single day to a second day (*ipso sequenti die Natalis Domini*) that supplements the first: what was decided on one day in October was ritually celebrated on one day in December. Operating along with the language of the calendar, and in fact making it possible, is the language of revelation: what provides the sole justification for precisely these two days and *only* these two days to be noticed in the same entry is their being moments of the same revelation (*manifesto Dei judicio*), and thus in virtually the same words the English reign is twice said to have ended because it is twice shown to have ended (*et Anglorum regno finem imposuit ... finitumque est regnum Anglorum*), once on the battlefield on one day and once at the coronation ceremony on another day. In the chronicle entry, these two events, a one-day war and a coronation ceremony, become two moments of one and the same singular event. Reading this way allows us to see that it is the very matter that ordinary source study wants to skip over – the assertions that God's intention has been revealed – that structure the *whole* entry: the assertion of divine intention is the precise determinant of how the chronicler sees the event as a coherent narrative structure with a beginning, a middle and an end. In other words, this element is what enables the chronicler to present two days separated by two months as a single event, that is to say, as both an epistemological and ontological unity – something he understands because that is how it happened in reality. The complex temporality of this singular event, in which something that happens later and in a different setting – in this case, the coronation ceremony in London – is considered as properly belonging to what happened at first, both completing it and in the process revealing or clarifying its original significance, is very familiar as typology to students of medieval biblical hermeneutics. For typology was the principal method of reading biblical narrative as a literal record of an event in its own time and simultaneously as a foreshadowing or prophecy of a second event which completes it and properly belongs to it, thus integrating contemporary Christian experience with its Hebrew past.[11] And this recourse on the part of the chronicler to typology is no mere accident or simple 'habit of thought', as old-fashioned history of ideas would have said. Rather, it is an intrinsic part of the chronicler's historical understanding of this event as a transfer of regimes.

William claimed the English throne first of all by descent, then by election and finally by feudal right.[12] From this point of view, the Norman Conquest, in standard textbook terminology, was not, as it were, a conquest, but both the recovery of what was rightfully already William's own while also being a continual manifestation of the justice of that recovery, a judgement of God in a large-scale aristocratic trial by combat over the rights of land tenure. In fact, Eadmer, writing at the end of the eleventh century, attributes this very way of interpreting the significance of the Norman victory to the French:

De quo proelio testantur adhuc Franci qui interfuerunt, quoniam, licet varius casus hinc inde extiterit, tamen tanta strages et fuga Normannorum fuit, ut victoria qua potiti sunt vere et absque dubio soli miraculo Dei ascribenda sit, qui puniendo per hanc iniquium periurii scelus Haroldi, ostendit se non Deum esse volentem iniquitatem.[13]

[Even now the French who were there say about this battle that although there was such various fortune on the one side and on the other, and nevertheless so many wounds and such flight on the side of the Normans, that their victory must truly and without doubt be entirely ascribed to a miracle of God, who in thus punishing the crime of Harold's perjury, shows that he is not a God who will allow iniquity.]

That a medieval chronicler signals this reading as a particularly French point of view indicates that there were others also in circulation. In noticing the way the chronicle entry presents the two days as a single event, we observed that the English reign is twice said to have ended in almost the same words (*et Anglorum regno finem imposuit ... finitumque est regnum Anglorum*). Let us note, too, that each day is rendered in its own single sentence. In its first sentence, the grammatical subject and agent of all the action is King Harold. In the precisely articulated Latin periodic syntax, everything is rendered as the attendant circumstances of three acts that King Harold performs: the first, his death, is subordinated to the latter two, which he performs, as it were, *post mortem*. Rushing to his death he puts an end both to the war and to the reign of the English (*rex Haroldus, corruens morte sua et bello et Anglorum regno finem imposuit*).[14] These are his last official acts as king, and they fully embody royal legitimacy: Harold, the last English king, brings *pax et iustitia* to his realm; he ends the war and it is he who in so doing ends English rule, 'by a manifest judgement of God'. If the first sentence is thus Harold's, the second sentence belongs entirely to William. In it, Count William becomes king (*Willelmus igitur Comes ... elevatus est in regem*). But in this, his only action, William properly does nothing: the main clause of the sentence is in the passive voice and the rest is made entirely of circumstantial and temporal constructions only loosely connected to him. The two sentences could not be more different. The first, Harold's sentence, is entirely controlled by the actions of the king; the second marks the precise moment when William becomes king and will be able to act henceforth. To this point, action (*potestas*) belongs to King Harold alone. The Hyde chronicler thus seems to take over the French perspective, that the conquest is in fact a divine judgement, but at the same time he maintains the legitimacy of Harold, as God's anointed king, to the last possible moment.

We need to observe one more thing before we are finished reading this little chronicle entry, something paradoxically most difficult to observe because it is so manifestly visible. The chronicle is written in Latin, and with Latinity inescapably comes a particular set of ways of rendering the social world, of framing experience and of asserting value. We have already observed that the chronicler uses the language of the calendar, but we did not say, among the great variety of possibilities that we can find in eleventh- and twelfth-century practice, that the chronicler uses the Roman method of counting.[15] Similarly, it is the language of ethnicity (*Normannos et Anglos*), of dominion (*regnum Anglorum ... regnum Normannorum*) and of imperium and territoriality (*factum est bellum in Anglia*) that makes this event part of a continuum of public affairs that begins in Rome and stretches without break to the contemporary world of the chronicler. Many other chronicle sources quite self-consciously use the Roman language of state, administration and sovereignty, and speak of what happened at Hastings as a transfer of regimes over a geographically and historically coherent territory. William of Poitiers, for example, deliberately and at length compares the Conqueror to Caesar at several points in his narrative, as does William of Malmesbury.[16] In the case of the Hyde Chronicle, we seem to be in the presence of an inescapable function of Latinity rather than a deliberate authorial choice, a function which is as important as a structure of signification as it is difficult to see: *gens, natio, princeps, regnum* and *respublica* are simply applied to the affairs of the eleventh century in the way that a Sallust or a Livy applied them to the public life of the first century. And yet, deliberate or not, the language that stresses the continuity of public affairs puts a deliberately regularizing inflection onto the transfer of regimes in 1066. The Hyde Chronicle's representation of events thus merges three otherwise separate versions of historical experience: one based on royal legitimacy and the efficacy of consecration, one based on aristocratic methods of judicial determination and a third on ancient Roman notions of the public sphere. And modern conditions of universal statehood make this third and very ancient version appear to be all the more naturally and unquestionably appropriate to the story being told and thus to the reality of the world being represented.

To 'see' all this as belonging properly to the chronicle's account of reality requires a reading that we can loosely call deconstructive. In the strict sense, deconstruction applies properly only to the philosophical work of Jacques Derrida, and particularly to his effort to understand the logic of Husserlian and Heideggerian metaphysics, a philosophical task that led him to an extensive and thorough meditation on the properties of writing.[17] In the weaker sense that I am using here, deconstruction refers in the first instance to a particular kind of critical reading devoted to understanding the operations that construct the text as a meaningful object: in order to see and understand them – in this case, the operations that allow the text to function as a document for the historian – the enabling devices of the text must be disassembled and isolated. This disassembly also requires that the reader take into account the interpretive acts that he makes not only in order to read in the particular way that he is reading (as a critical historian) but in order to read at all. For as we observed earlier in our discussion, the 'meaningfulness' of a text comes into being as a social fact, something that occurs in the encounter between a reader reading in a particular way for particular purposes and the written text produced in a particular way for its own purposes. Meaning is not simply poured into the reader by the writer through the medium of a text; it emerges rather at the intersection between the structuring activity of the text's language and the interpretive activity of the reader. Meaning is actively produced, not passively consumed.

We are accustomed to using metaphors of surface and depth to distinguish critical reading from what we think of as ordinary reading. We say that the critic discovers hidden meanings, reads between the lines or delves beneath the lines, finds buried implications, digs beneath the surface or mines the text. These metaphors are all misleading. There is nothing but white space between the lines and certainly nothing 'under' the surface. The fact, for example, that the Hyde Chronicle entry consists of two sentences, Harold being the active subject of the first and William the passive subject of the second, is as much a part of the literal surface of the text as is its assertion that the day before the Ides of October was a Saturday. What the reader – both the critical reader and his or her foil, the ordinary reader – does is traverse the surface of the text, and it is in that process of traversal that the signifying activity of the text operates. Part of my point here is that the kind of critical reading I have been pursuing leads neither to the dissolution of reality nor to a prison house of language with no exit. Rather, it allows us to see elements of reality embedded in the operations of language that otherwise we would not be able to see by 'ordinary' methods because it demands that we take all aspects of the text into account. Sentence structure is as meaningful in its way as information about who won the battle. In the case of the Hyde Chronicle, the matter of royal legitimacy seems to be of overriding concern: William did nothing in England before he was a consecrated king; as the chronicler rather vividly puts it, all the action was performed on a single day by God and his anointed representative, King Harold. And what of November and December? Medieval witnesses are surprisingly unforthcoming about the two months between the battle of Hastings and William's London coronation. For the Hyde chronicler those two months are precisely empty time: whatever happened was not only without significance, because the significance had already been revealed, but also was outside his normative modes of political understanding. There was no king in the land until the Christmas coronation. And without a king there was no public life.

In the relations that we have been discussing between a reader and a text, moreover, neither side is, strictly speaking, an individual. The reader comes to the Hyde Chronicle with all kinds of knowledge and expectations about reading, about chronicles, about the Middle Ages, about the work of the historian. This is an institutional knowledge born of previous reading, the encounter with previous texts, and in so far as it is disciplinarily informed, this knowledge does not belong only to the reader. The writer, too, comes to the task as already a reader. The writer thus produced this chronicle entry aware of such things as chronicles, prayers, sermons, theological treatises, confessional protocols and no doubt also of such things as romances, love poetry and lists of things to do. To put it directly, every reading takes place in the context of other reading; every writing takes place in the context of other writing; and every text makes its meaning intertextually, that is to say, in the context and subject to the influence of other texts.

Texts in contexts

Both literary critics and historians rely frequently on context as a control for interpretation. In speaking of the context, we generally mean to evoke a picture of something like 'life as it was lived by ordinary people in, say, the eleventh century'. In fact, we often use the expression 'the big picture' as the metaphor for precisely this kind of evocation of life. Where do we get this big picture from? Often, of course, it comes from the visual media of the period in question – our mental images of World War II, for example, are notoriously in black and white, at least until Steven Spielberg put them shockingly into colour in *Saving Private Ryan*. It comes as well, as the example of the Spielberg film indicates, from popular culture and certainly from more professionally respectable primary and secondary sources that we may have encountered already in our work. In our imagination of the Middle Ages, knights fighting dragons and rescuing damsels in distress jostle for position with architectural renderings of the weight-bearing elements of gothic cathedrals and techniques of royal administration. We very often hear historians typically accusing literary critics of falsely claiming that they are 'doing history' when in fact they are simply going to secondary sources, even superseded secondary sources, or worse yet, textbooks, to find instant backgrounds in which to 'contextualize' their reading. Literary critics, in turn, accuse historians of using complex works of literary art in a flat-footed manner as straightforward evidence for a social or intellectual background that they already know by other means. And art historians accuse everybody of using monuments and paintings as if they were merely pictures of life as it was actually lived and not complex symbolic objects in their own right. In each of these various cross-disciplinary accusations of unprofessional behaviour, the context is always being invoked as the stable, material ground in which to anchor the difficult, slippery and ambiguous meaning of a text. In these mutual accusations the context looks like what we already know – daily life in the eleventh century or the Norman Conquest, for example, or 'courtly love' or feudal society – and we situate the text, we say, within it or use the text rightly or wrongly as an illustration of it.[18]

The relation of text to context looks very different in the operations of traditional source analysis. When historians speak of 'doing history' they often mean to refer to a process of building up a context from the more or less fragmentary evidence at hand. In this activity, the source, a little piece of the past that we are attempting to reconstruct, is construed as a present part of a greater and lacking whole. There survives, for example, a record of a royal proclamation (what the specialists call a diploma) from the court of the Emperor Otto I, dating from the year 1001, that gives to the town of Cambrai the right to establish a market at Cateau-Cambrésis, and among other things to coin money there and institute officers for public affairs. This same diploma gives to the merchants at Cateau the same rights that 'the merchants at Cambrai' enjoy, and it says that any merchant who breaks the peace there will be subject to the same sanctions as are visited against the merchants of Cambrai. Historians have used this diploma, along with other very scanty surviving documentation, to develop a rather large claim about the expansion of the economy and changes in the population in northern France and Flanders in the early years of the tenth century. They have argued from the very existence of this diploma that the merchants of Cambrai and other ecclesiastical towns were already numerous enough by the year 1000 to deserve notice and even powerful enough to have their interests catered to by the emperor.[19] In this way, an argument about the surviving text is used to construct a large picture of the past of which the text is a small part. The ancient rhetoricians named this substitution of part for whole metonymy ('a meeting of the crowned heads of Europe' is a classic example), and indeed, the figure of metonymy is at the basis of much historical analysis: we treat a piece of evidence precisely metonymically when we fill in the patchwork provided directly by surviving evidence with the controlled inference of historical source study. In this process, it is the text which seems stable and material whereas the context is most definitely a construction: the text seems to be the solid evidentiary material out of which the necessarily more speculative context is constructed by the work of the historian.

In the act of thinking about a text in context the stable term thus moves from one side of the equation to the other depending on how we look at it. At one moment, the context seems to be the stable ground that limits the play of significances in the text; at the next moment, the text is the stable documentation that limits what can be said about its context. When one is stable, the other is unstable. Even brief reflection should convince us that since the location of stability depends on what we are doing, we need to conclude that in the dynamic act of contextualization there is in reality never a stable term. One powerful way of coming to grips with this indeterminacy is associated with the phenomenological approach of Hans Georg Gadamer, the so-called 'hermeneutic circle' in which one temporarily considers each side of the binary opposition, part and whole, to be 'known' or stable and moves successively from one side to the other, from using the whole to read the part and then from the part thus analysed back to the whole to reconstruct it, and so on.[20] In this approach, powerful as it is, an important aspect of the relationship between text and context is obscured, for text and context, part and whole, seem to be entities – and singular, material entities at that. In our previous discussion of evidence, I suggested that a text is rather a relation than an entity, or even more precisely considered, it is a set of relations – between production and reception, and above all between it and other texts that it both evokes and differentiates itself from. And if we reflect on the various examples of context in the previous paragraphs of this section (examples drawn from film and popular culture as well as from historical documents and secondary sources) we may be led to the inescapable conclusion that what we conventionally think of as 'the big picture' is itself inescapably composed of a variety of texts. We could go so far as to say that a text makes its own proper meaning precisely by marking a position among other texts that circulate within culture. In our reading of the Hyde Chronicle we saw it as a deployment of at least three perspectives – which we could conveniently name royalist, French and Roman – in its representation of the event of the Norman Conquest. And what are these perspectives composed of if not of texts?

What then do we mean by context? Fundamentally, the context of a text is a threefold set of other texts relevant to a particular act of reading. It consists, first, of those texts already circulating in culture at the moment of production of the text in question, texts that in various ways supply the writer of the text in question with a conceptual apparatus, a way of speaking and a provocation to write. It consists, too, of those texts of which the text in question takes direct account. For, as we have said before, writing takes place in a field already occupied by texts, and these 'pretexts' are cited, rewritten, avoided, dismissed and revised more or less overtly, more or less deliberately and more or less consciously by the writer in order to make the new text. They supply the writer with arguments to contend with, to agree with, to avoid or otherwise to take into account. In this way they could at any moment become present to the reader of the text. Any text is thus always inclusive of other texts out of which it is made and which in this way form its context. The great Russian literary analyst, M.M. Bakhtin, calls this relation of text to context 'dialogue', and has demonstrated in a brilliant series of readings that all texts are thus in internal dialogue with other texts.[21] In this way, the context is never something outside the text into which the text is placed; rather, in order to be a text at all, the text is permeated by other texts. Derrida's remark, 'Il n'y a pas de hors-texte' [there is no outside of the text], speaks of this directly. It sounds paradoxical, but it is nonetheless accurate to say that the context is thus already *inside* the text. Most obviously, the context is in the text because the text does not ever in any comprehensible way exist free of a context. There is no place to stand outside the text – even one as 'naïve' as our chronicle entry makes its meaning by deploying a finite but in actuality rather large series of other texts.

So far, I have treated context from the side of production. From the side of reception we can see the third textual relation of a text's context. For any individual reading is always informed by other readings. Our reading of the entry from the Hyde Chronicle, for example, is made possible by reading in biblical hermeneutics and medieval political theory as well as by other eleventh- and twelfth-century accounts of the death of King Harold. In this way, I have situated the Hyde text among a number of other texts that the chronicler may not have taken into account as such, but whose relevance is clear to my reading and to my sense of the past that I am attempting to understand. Some of these texts that inform my reading are contemporary with the Hyde Chronicle. Some are composed in other periods, among which I include immediately relevant secondary sources but also a great variety of primary and secondary materials that go to form my sense of the eleventh century. If seeing the context *within* the text requires an act that we can call deconstructive, then situating a text thus *within* a context requires an act of construction. The context is never simply given but comes into being by the very process of a situated reading.

The historian's work is thus both deconstructive and constructive, and unlike investigators in many other disciplines, historians must construct the very thing that they take as their object of knowledge, a particular segment of the past. This construction is not quite material reality, but rather a representation of material reality as the real object of historical knowledge. Michel Foucault's use of the term 'discursive formation' draws important attention to the linguistic substance of much of what we unreflectively take to be elements of material reality, a big picture of life in the past.[22]

The task of the historian

Jacques Derrida was not the first to notice that philosophers are first and foremost writers. What we actually name as Kantian philosophy is a collection of written texts with features conforming to the protocols of a particular genre. Similarly, in an article that he says he would have preferred to title 'Footnotes, Quotations, and Name-lists', J.H. Hexter discusses the stylistic characteristics that identify history writing as what it is – a paper to be read at a conference, for example, or an essay to be published in a professional journal such as *Medium Aevum*, a book for a university press or a class lecture – and distinguish it from the report of a physicist or a lyric poem.[23] Historians are readers and writers; what historians do takes place fully in writing.[24] They produce new texts by reading other texts in particular, professionally sanctioned ways and by writing in very particular, recognizable genres whose characteristics – including such things as footnotes, indices and pages of acknowledgements – serve to differentiate them from other similar texts in contemporary circulation, such as novels and historical romances. As a writer, the historian works constantly under the constraints of language on representation as such. The work of representing a reality that does not exist because it no longer exists is an act both of imagination and of literary composition, even for a historian working in the austere manner of a quantitative social scientist. The historian must secure the reader's assent to the likelihood that the historian's text is an adequate representation of the past. This compositional necessity allies the contemporary historian's work with that of the novelist, who also must secure the reader's assent to a reality constructed by the work of fiction. The novelist and the historian are both composers: they both must find convincing ways to sequence an event in language, to describe it with a thick enough texture of circumstance, to present the relation between individuals, social circumstance and changes over time in such a way as to gain the consent of their readership and their assent to the likeliness of the story being told. Now this similarity between fiction (a prose narrative that 'fictively' claims to be true) and historical writing (a prose narrative that claims a truth-telling intention) is unsettling, to say the least, and it has led to philosophical positions of extreme scepticism. The possibility of verifying any truth claim is compromised when we recognize that a truth claim is itself, strictly speaking, a performative utterance that takes place entirely within the confines of language, and the relation of words to the real world outside the text cannot be firmly demonstrated.[25] The metahistorical work of Hayden White and Louis Mink has carefully explored the implications of the compositional affinities between historical writing and fiction – White from a literary critical position of extreme scepticism and Mink from the point of view of classic philosophy of history.[26] Bernard Guenée, an important French medievalist, has used historiography to attempt to recover the possibility of source study as positive knowledge.[27] Gabrielle Spiegel has devoted much of her recent work to locating what she has disarmingly called a middle ground, theorizing the place from which it is possible to 'make' history without ignoring the unavoidable mediation of the discursive systems that always stand between the historian's utterance and anything that we might want to call 'the past'.[28] This is not the place to evaluate the relative merits of the various philosophical positions regarding the relation of textuality to historical work. What this multiplicity of perspectives demonstrates, however, is that textuality is an inescapable part of what the historian is faced with at every stage of work, from the analysis of sources to the creation of a finished argument. We cannot simply wish away our consciousness of language and its effects.

We live in a world saturated through and through with language and so did the people of the past. As we try to understand their world, both their moments of crisis and their daily life, we don't want to lose sight of the reality of their experience and their sufferings. The great fear provoked by the linguistic turn, that moment 'when everything became discourse',[29] to quote Jacques Derrida, was precisely the fear of the loss of contact with the humanity of the past. If everything was discourse, how would we find the world and understand the people in it? The answer seems to be given precisely by literary analysis. As we investigate the properties of representation, we discover that by taking into account not merely the things that are being said directly by our documents ('it was a Saturday', for example), but also the linguistic mechanisms that allow them to be said and said in the particular way that they are (that a single chronicle entry, for example, is composed of two sentences, one active and one passive), we discover that the reality we are engaged in understanding becomes thicker, less rarefied, more nuanced and multidimensional. And as we extend our inquiry outwards from the single source into examining the textual contexts and the intertextual play inseparable from the particular document on which we happen to be working, we uncover the continual social and cultural pressure on what is being said, how experience is being formulated, what is included and what is left out. The forms, connotations and even silences of our documents are as much a part of their linguistic surfaces as are their statements; the relation of a single text to the manifold network of texts of which it is a part belongs as surely to its properties as the language in which it is written; and all these things can be made to speak to us. The heavy weapons, as it were, invented to assault the complex textual objects of high culture have become admirable and useful tools for the construction of a past reality out of its fragmentary textual remains. If the linguistic turn threatened to pull the rug out from under working historians, it did so, paradoxically, to reveal the solid ground beneath their feet.

Guide to further reading

Erich Auerbach, *Mimesis: The Representation of Reality in Western Literature*, 50th Anniversary Edition, with a new introduction by Edward Said, trans. Willard R. Trask (Princeton, NJ, 2003).

M.M. Bakhtin, *The Dialogic Imagination: Four Essays*, trans. Caryl Emerson and Michael Holquist, ed. Michael Holquist, University of Texas Press Slavic Series no. 1 (Austin, TX, 1981).

Roland Barthes, *S/Z: An Essay*, trans. Richard Miller (New York, 1974).

Roland Barthes, *The Rustle of Language*, trans. Richard Howard (New York, 1986).

Jacques Derrida, *Writing and Difference*, trans. Alan Bass (Chicago, IL, 1978).

Jacques Derrida, *Of Grammatology*, trans. Gayatri Chakravorty Spivak (Baltimore, MD, 1998).

Michel Foucault, *Archaeology of Knowledge* (New York, 2002).

Edward Said, *The World, the Text, and the Critic* (Cambridge, MA, 1988).

Gabrielle Spiegel, *The Past as Text: The Theory and Practice of Medieval Historiography* (Baltimore, MD, 1997).

Paul Strohm, *Theory and the Premodern Text* (Minneapolis, MN, 2000).

Hayden V. White, *Metahistory: The Historical Imagination in Nineteenth-Century Europe* (Baltimore, MD, 1973).

Hayden V. White, *Tropics of Discourse: Essays in Cultural Criticism* (Baltimore, MD, 1978).

Hayden V. White, *The Content of the Form: Narrative Discourse and Historical Representation* (Baltimore, MD, 1987).

Notes

1   John E. Toews, 'Intellectual History after the Linguistic Turn: The Autonomy of Meaning and the Irreducibility of Experience', *American Historical Review* 92 (1987), pp. 879–907.
2   Toews, 'Intellectual History', pp. 898–9.
3   Although writing particularly about the meaning of a text, Paul Strohm defines the sense of meaning in general as something that comes into being as a historical operation very nicely: '"Meaning" is . . . always shared out or held in common. The meaning of a particular text exists somewhere in the range between broad tradition and unique articulation, between authorial intent and a broadened diversity of uses and appropriations, between the work's meaning to its intended and actual and subsequent audiences. Never unitary, a meaning's history and status (when? to whom? for what?) must always be specified'. Paul Strohm, *Theory and the Premodern Text* (Minneapolis, MN, 2000), p. xvi.
4   Toews makes a similar point: 'Intellectual History', p. 882.
5   Robin Fleming, *Kings and Lords in Conquest England* (Cambridge, 1991), for example, used the Domesday Book, intended by William I to be a survey of land tenure, as a means to reconstruct the family structure of much of England. Similarly, David Herlihy and Christiane Klapisch-Zuber used the tax records of the government of fifteenth-century Florence to get at large portions of Florentine public and private life, including such things as diet and marriage customs. See David Herlihy and Christiane Klapisch-Zuber, *Census and Property Survey of Florentine Domains and the City of Verona in Fifteenth Century Italy* (Madison, WI, 1977), machine-readable data file, updated 12 September 1999, available from http://dpls.dacc.wisc.edu/Catasto/index.html (accessed 2003).
6   Roland Barthes' 1971 essay, translated as 'From Work to Text', has been very influential in contemporary criticism. Barthes, too, insists on the

same relational quality – meaning comes into being in a complex negotiation that equally involves the side of reception (reading) as well as
production (writing) – and fully social implication that I have been trying
to present. 'The Text is that *social* space which leaves no language safe
outside, and no subject of the speech-act in a situation of judge,
master, analyst, confessor, decoder'. See Roland Barthes, 'From Work to
Text', in *The Rustle of Language,* trans. Richard Howard (New York,
1986), p. 64. Barthes here comes very close to a definition of Derrida's
famous and famously mistranslated remark, 'there is no outside to the
text (*il n'y a pas de hors-texte*)' that I will discuss below.

7   The terms are Antonia Gransden's, *Historical Writing in England* (Ithaca,
     NY, 1974), p. 411.

8   The text is from Edward Edwards (ed.), *Liber Monasterii de Hyda*, Rolls
     Series No. 45 (London, 1866), p. 294 (unless otherwise noted, all trans-
     lations are my own).

9   For a critical discussion of the eleventh-century sources for the death of
     King Harold, see Robert M. Stein, 'The Trouble with Harold: The
     Ideological Context of the *Vita Haroldi*', *New Medieval Literatures* 2
     (1998), pp. 181–204.

10   For a discussion of what is at stake in such picking and choosing, see
     Keith Hopkins, 'Rules of Evidence', review of Fergus Millar, *The Emperor
     in the Roman World (3rd BC–AD 337)* (London, 1977), *The Journal of
     Roman Studies* 68 (1978), pp. 178–186. The relation between plausibil-
     ity, probable argument and the establishment of truth has been seri-
     ously studied in western culture since at least the time of Aristotle. It is
     one of the central topics of forensic rhetoric, and yet despite the preva-
     lence of the forensic model and the common use of judicial metaphors
     in accounts of source study, the matter of plausibility has remained,
     with a very few exceptions, strikingly under-theorized in the philosophy
     of history. For an attempt to account for this under-theorization, see F.R.
     Ankersmit, *Cultural Memory in the Present: Historical Memory*
     (Stanford, CA, 2001).

11   See the classic studies by Jean Daniélou, *From Shadows to Reality;
     Studies in Biblical Typology of the Fathers*, trans. Wulstan Hibberd
     (London, 1960) and Erich Auerbach, 'Figura', in *Scenes from the Drama
     of European Literature* (Minneapolis, MN, 1984), pp. 11–78.

12   For a good summary of the grounds on which William rested his claim
     to the English throne, see Raymonde Foreville, *Histoire de Guillaume le
     Conquérant* (Paris, 1952), p. xvi.

13   Eadmer, *Historia Novorum in Anglia*, ed. Martin Rule, Rolls Series 81
     (London, 1884), pp. 8–9. One finds the language of divine judgement
     similarly used by Ralph of Coggeshalle, writing in England at the end of
     the twelfth century: 'Anno ab incarnatione Domini MLXVI. Willelmus
     dux Normannorum, contracto a partibus transmarinis innumerabili

exercitu, in Angliam applicuit apud Hastinghes, ac justo Dei judicio die Sancti Calixti Papae, regem Haraldum, qui imperium Angliae injuste usurpaverat, regno simul ac vita privavit'. See Ralph of Coggeshall, *Chronicon Anglicanum*, ed. Joseph Stevenson, Rolls Series 66 (London, 1875), p. 1.

14 *Impono* has a complex range of references, most of which are negative. Operative here is its sense of providing finality as in bringing an end to hope, putting the final stroke on a painting or bringing an end to war. See Charlton T. Lewis and Charles Short, *A Latin Dictionary* (Oxford, 1879, reprinted 1969), s.v. impono II. A.

15 Throughout the whole of the Middle Ages there is great variability in calendrical practice. Ranulph Higden's universal Chronicle, finished around 1340, for example, begins by numbering years from Abraham, then with the reign of King David starts over again, while also noting regnal years of Hebrew kings and years from the founding of Rome. For the period after Christ, Higden provides both Anno Domini dating and the regnal year of the Emperor of Rome to Charlemagne; afterwards he records the regnal year of any nation whose actions he discusses, and therefore primarily the regnal year of the English king. This multiplicity of dating in one chronicle is rather the rule than the exception. See Ranulf Higden *et al.*, *Polychronicon Ranulphi Higden Monachi Cestrensis; Together with the English Translations of John Trevisa and of an Unknown Writer of the Fifteenth Century*, eds C. Babington and J.R. Lumby, 9 vols, Rolls Series 41 (London, 1865).

16 William of Poitiers, *The Gesta Guillelmi of William of Poitiers*, trans. and eds R.H.C. Davis and Marjorie Chibnall, Oxford Medieval Texts (Oxford and New York, 1998). Cf. also William of Malmesbury, *Gesta Regum Anglorum*, eds. and trans. R.A.P. Mynors, R.M. Thomson and M. Winterbottom (2 vols, Oxford, 1998) , Book 3, section 254, subsection 3.

17 The degree of Derrida's importance is matched only by his legendary difficulty. The best entry point for Derridean deconstruction outside of the strictly philosophical camp are the essays collected in Jacques Derrida, *Writing and Difference*, trans. Alan Bass (Chicago, IL, 1978). The single most important of Derrida's texts for the field of literary analysis is Jacques Derrida, *Of Grammatology*, trans. Gayatri Chakravorty Spivak (Baltimore, MD, 1998). Perhaps the best discussion of Derrida's critical significance is Edward Said, 'Criticism between Culture and System', in *The World, the Text, and the Critic* (Cambridge, MA, 1988). The book as a whole is an invaluable contribution to literary analysis. For a discussion of Derrida's particular relevance to historical understanding, see Gabrielle M. Spiegel, 'Orations of the Dead; Silences of the Living', in *The Past as Text: The Theory and Practice of Medieval Historiography* (Baltimore, MD, 1997), pp. 29–43.

18 For a historian's reflections on the problems involved in invoking 'what we all know' to be the context, see Elizabeth A.R. Brown, 'The Tyranny of a Construct: Feudalism and the Historians of Medieval Europe', *American Historical Review* 79, no. 4 (1974), pp. 1063–88.

19 See Fernand Vercauteren, *Académie Royale de Belgique, Classe des Lettres et des Science Morales et Politiques, Memoires 33: Etude sur les civitates de la Belgique Seconde: Contribution à l'histoire urbaine du Nord de la France de la fin du IIIe au XIe siècle* (Brussels: Palais des Académies, 1934).

20 Hans Georg Gadamer, *Truth and Method*, trans. Joel Weinsheimer and Donald G. Marshall, 2nd rev. edn (New York, 1994).

21 See especially the essays collected in M.M. Bakhtin, *The Dialogic Imagination: Four Essays*, trans. Caryl Emerson and Michael Holquist, ed. Michael Holquist (Austin, TX, 1981).

22 See especially Michel Foucault, *Archaeology of Knowledge* (New York, 2002).

23 J.H. Hexter, 'The Rhetoric of History', in *History and Theory: Contemporary Readings*, eds Brian Fay, Philip Pomper and Richard T. Vann (Malden, MA, 1998).

24 The classical historian Averil Cameron describes the whole process very nicely: 'In order to write history – to generate a text – the historian must interpret existing texts (which will often be, but need not always be limited to, written materials, for ritual and social practice constitute texts too). But he will interpret, or "read" his texts in accordance with a set of other texts, which derive from the cultural code within which he works himself; and he will go on to write his text, that is, his history, against the background of and within the matrix of this larger cultural text. Thus history-writing is not a simple matter of sorting out "primary" and "secondary" sources; it is inextricably embedded in a mesh of text'. Averil Cameron, *History as Text: The Writing of Ancient History* (Chapel Hill, NC, 1989), pp. 4–5.

25 Nancy F. Partner, 'Making up Lost Time: Writing on the Writing of History', *Speculum* 61 (1986), pp. 90–117, reprinted in *History and Theory: Contemporary Readings*.

26 For Hayden White, see especially Hayden V. White, *The Content of the Form: Narrative Discourse and Historical Representation* (Baltimore, MD, 1987), Hayden V. White, *Metahistory: The Historical Imagination in Nineteenth-Century Europe* (Baltimore, MD, 1973), Hayden V. White, *Tropics of Discourse: Essays in Cultural Criticism* (Baltimore, MD, 1978); for Louis Mink see *inter alia* the essays by Louis O. Mink, in *Historical Understanding*, eds Brian Fay, Eugene O. Golob and Richard T. Vann (Ithaca, NY, 1987).

27 See the essays collected in Bernard Guenée, *Histoire et culture historique dans l'Occident médiéval, Collection Historique* (Paris, 1980),

and Bernard Guenée, *Politique et histoire au Moyen Age: Recueil d'articles sur l'histoire politique et l'historiographie médiévale (1956–1981)*, Publications de la Sorbonne, Serie Reimpressions no. 2 (Paris, 1981).

28 Gabrielle M. Spiegel, *The Past as Text: The Theory and Practice of Medieval Historiography* (Baltimore, MD, 1997).

29 'Structure, Sign, and Play in the Discourse of the Human Sciences' in *Writing and Difference*.

5

Finding the meaning of form: narrative in annals and chronicles

Sarah Foot

I shall use many voices in this history. Not for me the cool level tone of dispassionate narration. Perhaps I should write like the scribes of The Anglo-Saxon Chronicle, *saying in the same breath then an archbishop passed away, a synod was held, and fiery dragons were seen flying in the air. Why not after all? Beliefs are relative. Our connection with reality is always tenuous.*[1]

Annals are among the most familiar and perhaps the least problematical of the sources regularly encountered by historians of the European Middle Ages. In the simplicity of their form and the outward transparency of their contents, sets of annals look reassuringly straightforward and unambiguous. Once errors in transcribing dates and discrepancies between different manuscript traditions have been ironed out by modern editors, annals can reliably be pillaged for 'facts' (especially for names, dates and places) and conveniently cited in footnotes without further elaboration. In her musing on narrative forms, the fictional narrator of *Moon Tiger*, Claudia Hampton, quoted above, has conflated two eighth-century annals from the Anglo-Saxon Chronicle, but almost any single entry from the same part of the Chronicle would serve to reinforce her impression of the randomness of the contents. Take, for example, the record for the year 776: 'In this year a red cross appeared in the sky after sunset. And that year the Mercians and the people of Kent fought at Otford. And marvellous adders were seen in Sussex'.[2] The laconic understatement of this particular annal has attracted considerable scholarly attention, for the annalist inexplicably failed to report who had won at Otford, thereby depriving his audience of the one piece of evidence that would have given historical meaning to his narrative.[3] To the annalist, however, the natural phenomena observed in that year were apparently more noteworthy than the deeds of men against which he juxtaposed them. This example encapsulates the characteristics of the early medieval annal form: minimalist expression; the conjunction of outwardly unrelated observations in relentlessly paratactic style; an arrangement of discrete statements into a framework provided by the unbroken sequence of years numbered since the Incarnation. It is these features that explain the dubiety with which many modern historiographers have viewed this literary form.[4]

Conventionally, annals have not been seen as works of 'history'. A modern schematic understanding of the evolution of medieval historical forms places the annal at the earliest (primitive) developmental stage and the chronicle – a more expansive account of events similarly arranged within a chronological framework – at an intermediate phase, closer to the final, fully developed historical form, but still an imperfect form of history.[5] This neat categorization was not, however, made by early medieval Christian writers. Writing in the sixth century, Cassiodorus made a twofold distinction between chronicles, *chronica*, 'the mere shadows of histories and very brief reminders of the times', and the works of historians, those who 'recount the shifting movement of events and the unstable history of kingdoms with eloquent but very cautious splendour'.[6] In his *Etymologies*, the seventh-century Spanish bishop, Isidore of Seville, defined a different threefold hierarchy of historical writing, ranging from the calendar (which recounted the events of a month), through annals, descriptions of single years, to *historiae*, which are concerned with many years and times. History, he declared, is an account of contemporary events, while annals (*annales*) are about those years which our time does not know.[7] Isidore did not mention the chronicle in his discussion of historical forms (*de generibus historiae*); rather he placed it at the culmination of an ascending series of units of time (from an hour, a day, a week, a month, etc.) and used it as a means of describing and thereby explaining the Ages of the World.[8] Isidore's classification was adopted by the English historian Bede, who wrote in Northumbria in the early eighth century. Bede employed the genres of chronicle and history for wholly discrete purposes, incorporating an episodic world chronicle into his treatise on the reckoning of time (*De temporum ratione*) for the specifically theological purpose of demonstrating the continuity of divine providence over time, but treating the particular story of God's treatment of the English in connected narrative in his *Historia ecclesiastica gentis Anglorum*.[9]

It would thus seem that medieval writers differentiated writings about the past that were arranged chronologically from those that reshaped events with rhetorical skill to convey particular meanings. Chronicles (variously termed *chronicon, chronica, chronicae*)[10] and annals (*annales*) were together distinguished from texts that sought to offer more expansive or moralizing interpretations of the past, *historia*. Annals thus represented one form of chronicle; they did not constitute a discrete historiographical genre.[11] The twelfth-century writer, Gervase of Canterbury, made this point explicitly, drawing a distinction between *historiis* and *annalibus quae alio nomine chronica nuncupantur,* annals which others call chronicles,[12] but the two terms had been conflated long before the twelfth century.[13]

The conventional tripartite division between annal, chronicle and history is thus not particularly helpful; the prolix annal is hardly to be distinguished from the laconic chronicle entry, even if both were thought to serve a different literary (and moral) purpose from *historiae.* The question that remains open, and the issue which this essay seeks to explore, is whether the restricted form of chronologically organized medieval texts necessarily limited the uses to which such texts could be put. It asks whether it is possible that annalistic texts could ever provide a structured analysis of the past going beyond the flat recording of unedited 'facts' as they occurred.[14] Has the outward form of the early medieval annal, the year-by-year recording of events in continuing – potentially continuous – sequence, served to obscure hidden meanings within its content? Despite their organization in chronological sequence, might annals be considered narratives?

Narrative

When Lawrence Stone reflected on the revival of narrative in historical writing that he had observed in the 1970s, he defined a narrative as 'the organization of material into a chronologically sequential order and the focusing of the content into a single coherent story, albeit with sub-plots'.[15] A story confers both shape and meaning on events; it records not one thing *after* another ('the unmeaning "and next, and next, and next . . ." of reality'),[16] but one thing *because of* another. In the storyteller's mind a single central subject unites the separate statements, giving the tale both an overarching, organizing structure and a clear meaning linking the episodes narrated.[17] Stories have clear beginnings and ends. The narrative form, William Cronon has suggested, is 'intrinsically teleological' in that it explains how prior events or causes have led to a particular end; '[h]istorians and prophets share a common commitment to finding the meaning of endings.'[18] Hayden White has made a similar point: historical stories 'have a discernible form (even when that form is an image of a state of chaos) which marks off the events contained in them from the other events that might appear in a comprehensive chronicle of the years covered in their unfolding.'[19]

In outward form annals and chronicles look quite unlike narratives. They seem to provide only random assemblages of data, the raw record of events in sequence from which a historically more sophisticated mind might confect a story in the future, although that literary construction is not yet achieved.[20] The problem is neatly summarized by Robert Berkhofer:

> one-thing-after-another sequence is customarily labelled as an 'annal' or a 'chronicle', while a one-thing-because-of-another sequence is termed a

proper 'history'. The author or narrator connects the events and actions of the story through a plot and the action and events form a plot through a causal network of narration. Narrative, in short, constructs a context by connecting what seems unrelated into a story.[21]

Beyond chronological sequence, it is hard to find any underlying organizing principle behind a collection of annals from which to make their sequences of events intelligible. A chronicle may include the elements of a story or even obliquely seem to tell a story by implication, yet 'its form is not in itself meaningful'.[22] Constrained by the narrowness of the vision of its compiler(s), a 'mere' chronicle or set of annals does not provide a continuous or connected exposition, but rather offers only 'a conjunction of non-causal singular statements'.[23] Topolski reflected the general view of such texts when he argued: '[t]he annalist did not formulate any general concepts which would enable him to make a more expanded interpretation. He used terms that were commonly used and understood, and set the past in a conceptual framework which he himself created'.[24] A modern audience may struggle to recognize the significance and meaning of that conceptual framework. It is hard, as Cronon has argued, to establish the connections between recorded events, to establish their relative significance, or to keep track of their meaning: '[w]ithout some plot to organize the flow of events, everything becomes much harder – even impossible – to understand.'[25] Further, annals and chronicles are entirely open-ended. They lack inauguration, frequently simply beginning *in medias res*, and – more significantly – they lack closure, they fail to trace a sequence of events from start to finish. Since the arrow of time continues even after an annalist or chronicler has laid down his pen, both those genres are inevitably provisional; every passing year offers the potential for continuation so the story can never have end. It is largely because of this open-endedness that annals are customarily seen as lacking in narrative properties and thus as non-historical.

In an important contribution to narrative theory which attempted to break away from the notion of the imperfect historicity of annals and chronicles, Hayden White sought to illuminate the possible conceptions of historical reality that underpinned such forms of medieval historical writing. Yet he, too, took the view that the annal form 'lacks any narrative component' and argued that the chronicle could only aspire to narrativity. Part of White's argument is that the annalist (the example he chose was that of the Annals of St Gall, but the point is clearly to be taken more broadly to refer to all early medieval annals) is unable to transform his list of events into a discourse about the events considered as a totality evolving in time.[26] White found no connection between the events recorded, indeed no clear criterion for their selection; all the events included appeared to him to have the same order of importance or unimportance. Ultimately it was the absence of any plot connecting the annals that led White to see the form as unhistorical; the annalist's inability to impose meaning on the events that make up the story 'by revealing at the end a structure that was immanent in the events all along'.[27] For an account to be considered historical, White suggested, it is not enough for the events to be recorded in the order in which they originally occurred; the reality to which they bear witness can only be made meaningful if shaped in narrative form, with the formal coherency of a story imposed upon it.[28] Important and influential as White's analysis has been, it is limited by his insistence that the form and open-endedness of annals so constrains them as to make them ultimately incoherent, lacking moral meaning.[29] It is possible to argue beyond White's conclusions by reconsidering the assumptions that have underpinned conventional assessments of the historicity of early medieval annals and chronicles. If collections of annal entries are read not as discrete statements located only in time, but as unitary and coherent wholes, they can be shown to constitute more sophisticated analyses of the past conveying a larger meaning than has previously been recognized.

The earliest annals

Annals, compilations of brief notes of the significant events of a single year, are generally believed to have been written first in the margins of Easter tables, the tabulated record of the various astronomical data from which the date of Easter can be calculated for any given year. Blank spaces in the margins beside such a table offered opportunity not just for the recording of pieces of particularly important information, but also for fixing those events securely in time. Those responsible for compiling such records are sometimes credited with some historical awareness,[30] but 'the great mass of annalistic compilation' is generally seen as devoid of narrative components, or even of any strictly historical curiosity, because its content is restricted to the disconnected deeds of contemporary history.[31] The rigid temporal framework of the annal form, in listing the continuous (and continuing) sequence of the number of years that had elapsed since the Incarnation, inevitably imposed limits on its authors; only the briefest and most laconic statements of significant – or memorable – events might be fitted into such a grid.

The reckoning of time according to the Incarnation and the keeping of summary, dated records were conventionally thought to have emerged simultaneously, in which case the earliest annals must have been compiled in the seventh or early eighth centuries in the Christian West, after the circulation of the Easter tables of Dionysius Exiguus, in which years were numbered *ab incarnatione Domini*.[32] There is, however, evidence that annalistic notes were inserted into pre-Dionysiac Easter tables, into both the 84-year calendar used in Ireland and the 19-year cycle devised by Victorius of Aquitaine, which counted years since Christ's Passion.[33] To link the emergence of the annal form so closely with the adoption of dating from the Incarnation would seem mistaken, although an association between laconic record-keeping and paschal calculation is more persuasive.[34] More problematical is the question of whether the writing of more elaborate, chronologically organized notices or the writing of 'history' was significantly influenced by paschal annals.[35]

One of the attractions of the assumed connection between paschal tables and early annals is that the form of the Easter table – the tight control of the space on the manuscript page by the layout of the computistical material in columns – provides a plausible explanation for the distinctively laconic character of the content of annal entries. Each entry is necessarily constrained by the size of the blank space beside the year to which it relates. Since each new line signifies a new year, anything other than the briefest of remarks would flow confusingly to cover the space assigned to other years. What is less compelling is the traditional understanding of the development of early medieval historical forms that has seen the expansive annal as a mature form of the paschal annal and the chronicle as the former's direct descendant.[36] Arguing from the fact that most continental manuscripts containing Easter tables with annal entries post-date the earliest manuscripts of annal collections, Rosamond McKitterick has suggested that this understanding of the development of the form downgrades the annal too far. Her interest was in the production of annals at the Carolingian court, which she has located within a wider upsurge in interest in representing the past evident in the Carolingian period.[37] That this historical endeavour should properly be separated from the ahistorical activity of calculating in advance when Easter would fall is probably correct.[38] It is less obvious that Easter tables with annal entries represent an adaptation of the idea of developed annals, since such an argument ignores the writing of annals in paschal tables in the Gaelic world before Dionysiac reckoning was introduced into Ireland and also the late antique tradition of using Easter tables to carry historical information.[39]

One might more helpfully abandon the orthodox view that has represented the development of early medieval historical forms as an evolutionary process in which the annal is forever portrayed as a primitive specimen. We have already observed that it was the form of the Easter table that constrained and restricted the quantity – and possibly also the nature – of the material that could be recorded in the spaces beside each separate year. But a simple form should not automatically be assumed to reflect an unsophisticated or an unhistorical conception of the past. Nor ought the function for which a paschal table was originally copied (the charting of the liturgical *cursus* over a specified number of future years) be confused with the purposes such a table might serve once the Easters it recorded were past.[40] Rejection of the processual conception of the emergence of early historical forms opens the way to more subtle readings of annalistic texts, unconstrained by the presumption that these were in essence primordial organisms.[41] There is, of course, a qualitative as well as a quantitative difference between the laconic note in the margin of an Easter table (whether entered soon after the year to which it relates or at some much later date) and the much more expansive texts described as *annales* that were compiled in the eighth and ninth centuries, such as the Royal Frankish Annals and its continuations, the so-called Annals of St-Bertin and Annals of Fulda.

As a form of historical enterprise, the notes recorded in the margins of the Easter table most resemble the various sorts of lists compiled in the early Middle Ages, many of which (king-lists, lists of bishops or abbots, necrological notices, lists of landed donations and so on) might form the basis for discursive treatment in chronologically ordered texts or, alternatively, might have been excerpted from such texts.[42] Although limited by its form, a list has certain narrative characteristics, including chronological organization, a unifying central subject and, arguably, an elementary plot. A king-list, which records the names of kings in sequence and enumerates the length of each man's reign tells a story of the royal succession within a particular kingdom, possibly extending that story far back into a remote past.[43] That the 'plot' determining the nature of annal entries selected for retrospective inclusion in a paschal table is to us obscure does not necessarily mean that it was not entirely pellucid to the scribe who made the record (or to contemporary and even later medieval readers). Our ability to locate some logic in, for example, the recording of Roman consular succession in old Easter tables, should warn us against the characterization of seemingly inexplicable events of which notice is also made as 'random'.[44] We need to look more closely at some annalistic compilations and rethink these presumptions about the meanings that may lie within their disjointed form and laconic expression.

The form and style of annals

Doubts about the narrativity of chronologically organized early medieval texts are usually articulated in relation to their two defining characteristics: their outward form and literary style. Scribes often adopted a paratactic syntax, even when they were not constrained by tabulated space available on a manuscript page but were working within a framework which would permit them to provide as much information as they might choose beside the chronological marker for each separate year. In the mid-ninth-century Annals of St-Bertin, for example, events were simply recorded one after another, with minimal commentary. One part of the lengthy annal for 849 reads:

Charles [the Bald, West Frankish king 840–77] marched into Aquitaine. Nominoë the Breton, with his usual treachery, attacked Anjou and the surrounding district. The Northmen sacked and burned the city of Périgueux in Aquitaine, and returned unscathed to their ships. The Moors and Saracens sacked the Italian city of Luni, and without meeting the least resistance ravaged the whole coast along to Provence.[45]

With little here to indicate cause and effect, this reads as no more than a disjointed, disconnected sequence of barely related sentences and offers few of the characteristics of a narrative. Were one to attempt to read the whole of the account for each calendar year as a self-contained narrative, one would still struggle to find much by way of an overarching rationale determining the inclusion or exclusion of material. Like the text of which it is a continuation, the Royal Frankish Annals, the Annals of St-Bertin were centrally concerned with the deeds of the Carolingian family, predominantly but not exclusively with affairs relating to the West Frankish kingdom; each year's account reported where the king spent Easter and Christmas and the rest of the material appears to be organized in more or less chronological sequence within each period of twelve months.[46]

Entries in the Annals of St-Bertin and Annals of Fulda are far longer than those recorded in the vernacular English compilation, the Anglo-Saxon Chronicle, during the same period, but a similar parataxis characterizes those entries, too, with the conjunctives 'and' and 'then' used to link separate simple statements.[47] Perhaps the paratactic structure was intended in part to allow gaps for the reader to reflect on and understand the text. The pauses might have offered the reader or hearer time and opportunity to complete the narrative only sketched here. One should further recall that parataxis and apposition played an important role in Old English literary style and do not denote literary *naïveté*. The choice of the annalistic form in English or Frankish circumstances need not be seen as evidence for a lack of literary sophistication or artifice on the part of a compiler, nor as a reflection of the inability of those who constructed annals and chronicles either to devise alternative literary forms or to translate material presented via one medium into another. At one level it could be argued that the adoption of the annalistic form made some presentational problems easier: handling contemporaneous developments in separate English kingdoms or across different parts of the Carolingian Empire without recourse to digression or recapitulation was achieved more readily if events were set against the chronological sequence of dates than if they were reported in any other form. Indeed the notion of simultaneity could be thought to be important to the central argument of these works read as a whole.

Questions of stylistic preference cannot be separated from the issue of the spatial constraints placed on those who sought to insert historical notes beside a pre-existing tabular list of dates. Not only, as we have seen, can it be difficult to discern any coherent theme linking laconic statements, but individual notes are frequently disjointed. Blank years (sometimes several sequential years) can separate textual entries, where a date, the reference to the period of time elapsed since the Incarnation, has been counted but no historical information recorded. In instances where historical notes seem to have been added to a tabulated list of dates compiled for another purpose, the years without annotation are perhaps more readily to be understood than they are in manuscripts which have disaggregated annalistic records from any liturgical or other material with which they might once have been associated. In the latter case, it is hard to know quite how to interpret these blanks or to account for their having been recopied.[48] Were these years when nothing noteworthy happened or periods for which no one was able to remember anything that had occurred? If so, were the empty years copied – and recopied – in the expectation that one day the lacunae could be filled?

Two possible answers present themselves. One could take the scattering of 'random' jottings across uneven intervals of time as indicative of the ahistoricity of the enterprise of collecting and copying annals. Alternatively, one might see the chronological framework as the central organizing principle of the annal form, a conception of time that determined not only the appearance and layout of the material on the page but also its content.[49] The sequence of dates could be conceived as a linear representation of all the years elapsed since the Incarnation, a line tracing the natural, divinely ordained progression of historical time from the birth of Christ, through the recorded past and on, beyond the present of the scribe, into an infinite future. On this reading, the annalistic record would serve to locate the past of the people or community whose deeds it recorded within the same divine temporal framework and thus within God's plan for redeemed humanity. The temporal conceit here becomes central to the entire enterprise. Far more than simply providing a schematic grid into which data could be inserted as appropriate or as available, the counted years convey an intrinsic story to which the events and deeds of men noted by the scribe are only a gloss. The omission of even one year would be unthinkable: God's time is sacred and therefore irreducible. One simple but effective schema could thus be adapted to demonstrate the evolution of the English, the emergence and success of the Carolingian dynasty, the fate of the kingdoms east of the Rhine or the preoccupations of a northern Frankish monastic community. If conceived in this way then both short, laconic annal entries and more expansive accounts of sequences of years do far more than measure the intervals between a series of limited, abstract instants. By writing within this genre, the compilers have contrived to colonize a concrete temporal space, locating themselves and the subjects of their texts into the larger story that is time itself. They have woven complex and sophisticated pictures, plotting – as if on a graph – the moments at which different means of counting time intersect, above all the coincidence of divine and human time.[50] Annals '[impose] syntax on time':[51] they should properly be considered as narratives.

Plotting time

The outward form of annals and chronicles can, as we have just seen, be interpreted not as a limiting characteristic reflecting an unhistorical conception of the past, but rather as a carefully contrived and skilfully manipulated mode of presenting a particular story within a universal frame. Annalists make significant patterns from sequences of events; if sets of annals are read not as the sum of their separate parts but as unitary texts, they convey narratives. There is a continual tension between the syntactical parataxis of the record of each separate year and the rhetorical unity of intention that characterizes the whole, a tension which the reader must transcend before the wider meaning will become clear.[52] But once these texts are read entire, the relationship between the separate – outwardly random, disjointed and unconnected – annals becomes manifest. In each of the cases explored here (and arguably in other early medieval annal collections also), the distinct strands have been deliberately selected by their compilers to construct a meaningful plot, a dynamic sequential skeleton, of which chronological sequence is a significant organizing principle but not the sole determinant of the selection of material for inclusion.

Few sets of annals survive to us in a form compiled and written by only one author; most of these texts evolved accretionally as several compilers and subsequent revisers reworked earlier material or carried the story forward into their own time. In this sense they may be, as Liesbeth van Houts has suggested, 'histories without an end'.[53] Manifestly, of all those involved in such a collaborative enterprise it is the last reviser or continuer of an annalistic compilation who speaks most clearly to a modern audience reading his manuscript. All responsible, however, may reasonably be assumed to have shared many, if not all, of the aspirations of the earliest redactors of the text; otherwise, why did each elect to revise and extend his predecessors' efforts and not compose an entirely new text of his own? One possible reason why these texts were so often continued may be that these compilations told a story in which subsequent generations felt themselves to be participants also. Not only could that story perpetually be referred back to the beginning of Christian time and into the historical world of the Old Testament, but equally it strained to look forwards to the second coming and the end of time as men knew it. The cosmology within which annalists and chroniclers wrote had a clear, well-known form and direction, even if the texts which they produced appear to lack both. This wider point may be illustrated from two examples.

The Royal Frankish Annals

The Royal Frankish Annals provide a description of Frankish history from 741 to 829, arranged as year-by-year accounts but probably not written in the form in which they have survived (nor in their revised version) on an annual basis at Charlemagne's court.[54] The annals tell a story with a complex and sophisticated message that focuses on the Franks as a *gens*, a people with a shared destiny, in a way that no earlier writers had done, and they do so by closely associating the Franks' distinct identity with the Carolingian family.[55] No distant past is reported here: that past had been colonized by the Merovingian dynasty, whose deposition by Pippin (son of Charles Martel and, like his father, mayor of the royal palace) the annalist describes so casually under the year 750.[56] Rather the Royal Frankish Annals tell a story that has a clear beginning. The sequence of annals opens in the year 741 with the laconic statement: 'Carolus maior domus defunctus est' (Charles mayor of the palace died).[57] Only a reading of the later annals – or a prior knowledge of the subsequent development of Frankish history – would show the reader how the death of Charles Martel inaugurated a new era in the history of the Frankish people. That understanding emerges gradually as the narrative develops through these annals, which, far from merely 'ordering disorder',[58] were carefully designed to present a particular story creating the Franks as a people chosen by God and specifically linking their destiny with the Carolingian family, first as mayors of the palace and then as kings. The central subject of this story was the king and his deeds, performed 'in concert with, with the consent of, with the support of the Franks'.[59] Because those acts did not naturally organize themselves within the span of a calendar year, nor did the narrative, which frequently flowed across the artificial break of the change of date. For example, the annal for 780 ends with the statement that King Charles decided to go to Rome with his wife to pray there and that, en route, 'he celebrated Christmas in the city of Pavia, and the date changed to 781.' The new year number is little more than punctuation here, for the narrative continues, 'After resuming this journey he celebrated Easter at Rome'.[60] The syntactical flow of the story across the year end (and the failure to name the subject of the pronoun 'he') argues strongly against the possibility that the accounts of each year should be read as discrete narratives. The year numbers are not insignificant, however, for they fix the Frankish people and their kings into the linear progression of Christian time, and the constant report of where the king spent Easter and Christmas each year further locates the rhythms of the secular court into the cyclical revolution of the liturgical calendar.[61] Time drives the narrative relentlessly forward through various moments of crisis (the abortive Spanish expedition of 778), triumph (the imperial coronation recorded under the year 801) or curiosity (the arrival in 802 at Aachen of the elephant called Abul Abaz, or the remarkable astronomical phenomena reported in 807). What the Royal Frankish Annals lack, as do other sets of early medieval annals, is a clear end. The text stops abruptly in mid-tale with an account of how the Emperor Louis (the Pious, only surviving son of Charlemagne) spent the winter of 829 at Aachen, celebrating there the feasts of Martinmas (8 November), the holy apostle Andrew (30 November) and Holy Christmas 'with much joy and exultation'.

That this story was far from completed is clear from the fact that it was continued in other sets of annals. The Annals of St-Bertin take up the story where the Royal Frankish Annals left off – 'In February, an assembly was held there at which he decided to undertake a campaign with all the Franks into the lands of Brittany' – providing no new heading, nor even any identification of the place (Aachen) or naming the 'he' (Louis) who made the decision to campaign.[62] The interruption in the annals can be explained by the events of 830, when a rebellion against the emperor led to the dispersal of his entourage, including the arch-chaplain Hilduin who was probably responsible for the annals' compilation. Their continuation, first at the royal court and later in other centres, witnesses powerfully to the extent to which the story central to the Royal Frankish Annals continued to resonate to an audience in the 830s. At its heart is the story of the collective memory of the people under Frankish rule. The annals offer a narrative of the way the Carolingians saw their past, telling that family's story as part of the larger story that is the history of Franks.[63]

The Anglo-Saxon Chronicle

A similar argument can be advanced for the compilation of Old English annals known as the Anglo-Saxon Chronicle. Despite their unpromising form (and rigid, tabular manuscript layout),[64] the text read as a whole can be shown to be more exegetical than its separate parts may outwardly appear. I have argued elsewhere that the common stock of the Anglo-Saxon Chronicle, the first recension of annals running from 60 BC to 891, must be understood within the context of vernacular writing produced at the court of King Alfred. Although it seems unlikely that the king took a personal role in the construction of this text, his ideas about the history shared by all his newly united people, the *Angelcynn*, pervade the text in conception and execution.[65] The Chronicle plots, as a series of parallel tracks, a selection of events known and available to the compilers which recount the pasts of all of the separate English kingdoms within one overarching story, the story of the making of the *Angelcynn*. There is no clear authorial voice, for differences in vocabulary distinguish some groups of annals, suggesting that they were originally compiled in different places and arguably at different times, yet there does seem to be a unifying authorial intent. Individual voices have not been ironed out in the collaborative process of compilation, but all those voices are telling elements of the same story, the story of the English. Each of the past events, persons and phenomena with which that tale is concerned has been endowed with identity by being plotted into an overarching story which confers retrospective meaning on episodes that, unplotted, might seem incomprehensible.

Oddly, and in marked contrast to the Royal Frankish Annals, there is no single beginning to this story. Rather, the English elements of the Chronicle have multiple beginnings. The Anglo-Saxon chroniclers, unconstrained by the political difficulties that had beset the Carolingian annalists, expressed an interest in the distant past; they located the story of the island of Britain within the history of the Roman Empire by recording the date of Julius Caesar's arrival on its shores and noting various events relating to the history of Roman and sub-Roman Britain, and others taken from various universal histories that fell in the early years of the Christian Church. Yet once the 'Anglo-Saxon' story begins, the tale starts over several times, for the arrival of each separate English people is recorded as a new beginning. These are among the most difficult portions of the Chronicle to interpret, with their mingling of history and myth, their duplication and repetition. Although there is a preponderance of West Saxon material in these early years, this is not an exclusively West Saxon chronicle; other royal pedigrees as well as the West Saxon ones are included, Northumbrian, Mercian and Kentish.[66] Furthermore, the genealogical material explicitly links the stories of present rulers to those of both the mythical past of the migration period, but further to the 'historical' past of the Old Testament, in tracing a lineage for these kings to Woden, and then backwards from Woden to Christ.[67]

Once the separate migrant groups have coalesced into recognizable kingdoms under royal authority, the story starts to flow with greater fluency, but without compromising its multi-textured nature. West Saxon events are juxtaposed with the stories of the peoples of Kent, Mercia, East Anglia and (to a lesser extent) Northumbria, their political development and, in each case, the moment of their conversion to Christianity. A significant sub-plot emerges in this narrative in the late eighth century: the attack and subsequent invasion of the Anglo-Saxon kingdoms by a new pagan outside people, and the fate of each separate kingdom in response. Here the West Saxon victory is celebrated, in stark contrast to the failures of the other native English royal lines, but it is explained as a victory for a whole people (and for the continuance of their shared faith), not just for one dynasty. This, at least in part, answers what may seem to be the most irresolvable paradox about the content of the Chronicle, namely the function of the non-West Saxon (and western Mercian) material. It is not so much that the inclusion of the stories of the other peoples celebrates their achievements, as that it shows that their separate histories, their small triumphs and ultimate failures, were a necessary step in the divine plan. In the 880s, the Anglo-Saxon Chronicle records laconically, but with a certain relentless inevitability, the gradual disintegration of the Carolingian world in western Europe. Perhaps the Frankish elements are included here in the narrative in order to set Alfred's newly formed English people in a wider European perspective, to demonstrate the contrast between the unification of the English and the disintegration of the Franks. In making direct comparison with the responses articulated by those who had experienced the same onslaught from pagans from the north, the chronicler(s) illustrate(s) Alfred's superior capacity in wisdom and in warfare over his continental counterparts. Northern and eastern England may have fallen under Danish, pagan rule, but the men of Kent and of western Mercia, who had sworn their loyalty to Alfred in the general submission of 886, were being left in no doubt that this was a good leader to follow.

The end to which this narrative was directed was the linking of the separate strands, those Christian Anglo-Saxons outside the Danish areas, into one united people. But this was not the end, of course, not merely because the sequence of annals continued beyond 886, but also because implicit in the thinking behind the creation of the Chronicle is the notion that this people, the *Angelcynn*, created out of the political necessities imposed by the Danish wars, would have a future together as one people – a future to which all the past here accumulated had been directed.[68] The historical models provided by the salvation histories of Augustine and Orosius are clearly pertinent (and were obviously available, since these were among the works 'most necessary for all men to know', that King Alfred had included in his translation programme). Yet there is an important sense in which the Anglo-Saxon Chronicle can be shown to function as a specifically Christian history, because it was necessarily concerned with working out God's intentions for the English; it is prophetic in both its structure and its goals.[69] The Chronicle thus fits well within the available salvationist historical models and within Augustinian notions of the ages of the world. According to the story told by the Chronicle, by the time of its compilation, the English have reached the early manhood of the fourth age, equivalent in biblical terms to the period from the time of David to the Babylonian captivity.[70] Alfred's rule can thus be understood as restoring the English to life under the law, bringing the English to maturity, to later manhood and so bringing them into the fifth age.

The Anglo-Saxon Chronicle was very much a product of the political realities of the later ninth century. Although at one level (like all sets of annals) the Chronicle could tell only a provisional tale, it does self-consciously point towards a future beyond the continuous present to which the narrative had so far been brought. In that Augustinian sense, the accumulated English pasts prefigured what was to come; they were 'the shadow of the future',[71] a future in which the English as a chosen nation would surely be glorious under their West Saxon kings, even if the tale told so far fell short of that hope. There are points in the continuations of the Chronicle when this notion seems to be remembered – perhaps in the annal for 896 or in Alfred's obituary recorded under the year 900, certainly in the 937 Brunanburh poem:

> Never yet in this island before this by what books tell us and our ancient sages, was a greater slaughter of a host made by the edge of the sword, since

the Angles and Saxons came hither from the east, invading Britain over the broad seas, and the proud assailants, warriors eager for glory, overcame the Britons and won a country.[72]

Only God knows the end, and the end of 'English' history will be the end of time itself.

The meaning of form

I have argued here that the form of annals and chronicles is not an impediment to comprehension but is a central element in conferring meaning on their content. If sets of annals are read entire, rather than as random assortments of variously collected (and unedited) notes, they convey significant narratives. Annals are not mere recitations of everything that happened within a given time-span, but the self-conscious construction (emplotment) of cogent stories, made meaningful by selection, omission and careful interpretation. There is matter that is appropriate for inclusion in texts of this genre – the deeds of kings and acts of bishops, war, plague, famine and astronomical phenomena – and there is matter which is not, and which is not found in these court-produced annals or in monastic collections – Brother Martin's annual haircut, or Sister Ælfgifu's night of passion with the marauding viking Olaf are equally inappropriate and both properly ignored. The Royal Frankish Annals and their continuations and the Anglo-Saxon Chronicle are multi-textured stories; at one level each tells an account of one victorious dynasty (with particular emphasis on the triumphal military deeds of key individuals in that family) and displays the right of that lineage to rule a single people. But these texts do more than offer propaganda for a single royal line, for each provides the history of a people. The Anglo-Saxon Chronicle colonizes a defined space in the remote past and shows, by tracing the parallel tracks of each separate English people's development how it was that they were ultimately brought under one man's rule. The Royal Frankish Annals, as Rosamond McKitterick has shown, 'forge a Frankish identity by constant reiteration and triumphal narrative'.[73] Both consolidate their stories within a Christian chronological framework, made explicit through the recitation of the linear sequence of elapsed time. The meaning of the form is Christological: annals plot human endeavour within a framework of God's devising. Only God knows the secret of the 'blank' years and only He can plot the moment when time will end.

> This Age likewise, which is now running its course, will also have a duration uncertain to mortal men but known to Him alone who commanded his servants to keep watch with loins girded and lamps alight.[74]

Guide to further reading

For a general introduction to the question of annals, chronicles and medieval history see:

Dumville, David, 'What is a Chronicle?', in Erik Kooper (ed.), *The Medieval Chronicle II* (Amsterdam and New York, 2002), pp. 1–27.

Hay, Denys, *Annalists and Historians: Western Historiography from the Eighth to the Eighteenth Centuries* (London and New York, 1977).

McCormick, Michael, *Les annales du haut moyen âge* (Turnhout, 1975).

Poole, R.L., *Chronicles and Annals: A Brief Outline of their Origin and Growth* (Oxford, 1926).

For analysis of particular groups of annals see:

Clark, Cicely, 'The Narrative Mode of *The Anglo-Saxon Chronicle* before the Conquest', in Peter Clemoes and Kathleen Hughes (eds), *England before the Conquest* (Cambridge, 1971), pp. 215–21.

Collins, Roger, 'The "Reviser" Revisited', in Alexander Callander Murray (ed.), *After Rome's Fall: Narrators and Sources of Early Medieval History* (Toronto, Buffalo and London, 1998), pp. 191–213 (on the revision of the Royal Frankish Annals).

Hen, Yitzhak, 'The Annals of Metz and the Merovingian past', in Yitzhak Hen and Matthew Innes (eds), *The Uses of the Past in the Early Middle Ages* (Cambridge, 2000), pp. 175–90.

Nelson, Janet, 'The Annals of St-Bertin', in Margaret T. Gibson and Janet L. Nelson (eds), *Charles the Bald: Court and Kingdom* (2nd rev. edn, London, 1990), pp. 23–40.

For a wider discussion of early medieval uses of the past see:

Foot, Sarah, 'The Making of *Angelcynn*: English Identity before the Norman Conquest', *Transactions of the Royal Historical Society*, 6th series, 6 (1996), pp. 25–49.

McKitterick, Rosamond, 'Constructing the Past in the Early Middle Ages', *Transactions of the Royal Historical Society*, 6th series, 7 (1997), pp. 101–29.

Notes

1 Penelope Lively, *Moon Tiger* (London, 1987; paperback edn, 1988), p. 8.

2 Anglo-Saxon Chronicle, 776. All citations from the Chronicle are given, with corrected (not manuscript) dates, from the translation by Dorothy Whitelock, published in *English Historical Documents I: c.550–1042* (2nd edn, London, 1979), no. 1.

3 Charter evidence reveals that the men of Kent overthrew their Mercian overlords at Otford, for their kings reverted from 776 to their former habit of granting land without reference to any Mercian overlord until, c.785, the Mercian king, Offa, reasserted his authority over Kent and ruled the kingdom directly from then until his death in 796: S. Keynes, 'Changing Faces: Offa, King of Mercia', *History Today* 40 (November 1990), pp. 14–19.

4 Some have even argued that the form is not literary: Antonia Gransden, *Historical Writing in England c.550–c.1307* (London, 1974), p. 29.

5 R.L. Poole, *Chronicles and Annals: A Brief Outline of their Origin and Growth* (Oxford, 1926); C.W. Jones, *Saints Lives and Chronicles in Early England* (Ithaca, NY, 1947), pp. 26 and 34; Michael McCormick, *Les annales du haut moyen âge* (Turnhout, 1975).

6 Cassiodorus, *Institutes*, ed. R.A.B. Mynors (Oxford, 1937), I.17.1–2, pp. 55–7; quoted by Steven Mulberger, *The Fifth-Century Chroniclers: Prosper, Hydatius, and the Gallic Chronicler of 452* (Leeds, 1990), p. 8.

7 Isidore, *Etymologiae*, ed. W.M. Lindsay (2 vols, Oxford, 1911), 1, 44.4: 'Inter historiam autem et annales hoc interest, quod historia est eorum temporum quae vidimus, annales vero sunt eorum annorum quos aetas nostra non novit', quoted by Bernard Guenée, 'Histoires, annales, chroniques. Essai sur les genres historiques au Moyen Age', *Annales: Economies, sociétés, civilisations* 28 (1973), p. 1001, n. 27.

8 This is logical when one recalls that the root of the Latin word *chronica* is the Greek noun χρονος. Eusebius of Caesarea had described his 'Chronicle' as 'tables about time', χρονῖκοῖ κανόνες: David Dumville, 'What is a Chronicle?', in Erik Kooper (ed.), *The Medieval Chronicle II* (Amsterdam and New York, 2002), p. 1. Isidore inserted a world chronicle in the fifth book of his *Etymologiae*, V. 29. See Faith Wallis, *Bede: The Reckoning of Time* (Liverpool, 1999), pp. lxvii–lxix and 353–5.

9 Wallis, *Bede*, pp. lxx–lxxi.

10 Dumville, 'What is a Chronicle?', p. 2.

11 Dumville, 'What is a Chronicle?', p. 4. Compare Kenneth Harrison, *The Framework of History to A.D.900* (Cambridge, 1976), p. 50: 'If a line between annals and chronicles cannot clearly be drawn, neither can the transition to history'.

12 *Cronica* of Gervase of Canterbury, ed. William Stubbs (2 vols, London, 1879–80), prologue, vol. 1, p. 87.

13 Contra Poole, *Chronicles*, pp. 33–4: short annals represent the primitive form; large annals 'belong to a later time, when Annals were developing into Chronicles and were half-way on the road to becoming Histories'. And also Bernard Guenée, *Histoire et culture historique dans l'Occident médiéval* (Paris, 1980), pp. 203–4.

14 Jerzy Topolski, 'Historical Narrative: Towards a Coherent Structure', *History and Theory*, Beiheft 26 (1987), p. 78.

15 Lawrence Stone, 'The Revival of Narrative: Reflections on a New Old History', *Past and Present* 85 (1979), p. 3.

16 Nancy Partner, 'Making up Lost Time: Writing on the Writing of History', *Speculum* 61 (1986), p. 94.

17 W.H. Dray, 'On the Nature and Role of Narrative in History', in Geoffrey Roberts (ed.), *The History and Narrative Reader* (London, 2001), p. 28.

18 William Cronon, 'A Place for Stories: Nature, History and Narrative', *Journal of American History* 78 (1992), p. 1370.

19 Hayden White, *Metahistory: The Historical Imagination in Nineteenth-Century Europe* (Baltimore, MD, 1973), p. 6. Compare his *The Content of the Form: Narrative Discourse and Historical Representation* (Baltimore, MD, 1987), p. 21: 'for the want of [closure] the chronicle form is adjudged to be deficient as a narrative'.

20 White, *The Content*, p. 5.

21 Robert F. Berkhofer, *Beyond the Great Story: History as Text and Discourse* (Cambridge, MA, and London, 1995), p. 37.

22 M.C. Lemon, 'The Structure of Narrative', in Roberts (ed.), *The History and Narrative Reader*, p. 109; White, *The Tropics of Discourse*, p. 109.

23 Morton White, *Foundations of Historical Knowledge* (New York and London, 1965), p. 4.

24 Topolski, 'Historical Narrative', p. 78.

25 Cronon, 'A Place for Stories', p. 1351.

26 White, *The Content*, pp. 5 and 16.

27 White, *The Content*, pp. 7 and 20.

28 White, *The Content*, pp. 20 and 42.

29 White, *The Content*, p. 21.

30 R.W. Southern, 'Aspects of the European Tradition of Historical Writing 1: The Classical Tradition from Einhard to Geoffrey of Monmouth', *Transactions of the Royal Historical Society*, 5th series, 20 (1970), p. 180.

31 McCormick, *Les annales*, p. 13.

32 Poole, *Chronicles and Annals*, p. 26; McCormick, *Les annales*, p. 27; Guenée, *Histoire et culture*, p. 203.

33 Harrison, *The Framework*, pp. 32–3 and 45–6; Daibhi O'Croinin, 'Early Irish Annals from Easter Tables: a Case Restated', *Peritia* 2 (1983), pp. 74–86.

34   Dumville, 'What is a Chronicle?', p. 7.
35   Harrison, *The Framework*, p. 45. For a maximal view see Jones, *Saints Lives*, p. 116: 'the introduction of such a list [as an Easter Table, or list of moveable feasts] . . . eventually changed the whole course of historical writing'.
36   This point was made by Harrison, *The Framework*, p. 45.
37   R. McKitterick, 'Constructing the Past in the Early Middle Ages', *Transactions of the Royal Historical Society*, 6th series, 7 (1997), p. 113. Also Matthew Innes and Rosamond McKitterick, 'The Writing of History', in Rosamond McKitterick (ed.), *Carolingian Culture: Emulation and Innovation* (Cambridge, 1994), pp. 193–220.
38   McKitterick, 'Constructing the Past', p. 113.
39   Dumville, 'What is a Chronicle?', p. 7.
40   Contrast here readings of McKitterick, who thinks it ironical 'that tables designed to set out both the rhythm for the future and chart the cyclical liturgical year should have given rise to notes about the past' ('Constructing the Past', p. 111) and Dumville, who has stressed the fact that there are sufficient surviving examples of historical information being entered retrospectively across larger groups of tables for it to be clear that Easter tables were seen as bearers of systematic historical record as well as notices of random events ('What is a Chronicle?', p. 9).
41   Compare White, *The Content*, pp. 5–6.
42   Dumville, 'What is a Chronicle?', pp. 9–11.
43   See, for example, Molly Miller, 'The Disputed Historical Horizon of the Pictish King-lists', *Scottish Historical Review* 58 (1979), pp. 1–34.
44   Compare Dumville, 'What is a Chronicle?', p. 9.
45   *The Annals of St-Bertin*, *s.a.* 849, translated by Janet L. Nelson (Manchester, 1991), p. 68.
46   Janet L. Nelson, 'The Annals of St-Bertin', in Margaret T. Gibson and Janet L. Nelson (eds), *Charles the Bald: Court and Kingdom* (2nd rev. edn, London, 1990), pp. 23–40.
47   Cicely Clark, 'The Narrative Mode of *The Anglo-Saxon Chronicle* before the Conquest', in Peter Clemoes and Kathleen Hughes (eds), *England before the Conquest* (Cambridge, 1971), pp. 215–21.
48   Consider, for example, the Parker, 'A', manuscript of the Anglo-Saxon Chronicle: Cambridge, Corpus Christi College (hereafter CCCC), MS 73, fo. 3v. The dates for the years AD 262–345 are laid out here in two columns, but the sole entry beside any of these years (for AD 283, recording the martyrdom of St Alban) is a later addition made to the manuscript at Canterbury. There is a facsimile of this manuscript, eds R. Flower and H. Smith, *The Parker Chronicle and Laws* (London, 1941).
49   White has also explored this possibility: *The Content*, pp. 8–9.
50   I intend to return to the question of the conceptions of time in the Anglo-Saxon Chronicle on a future occasion.

51 Partner, 'Making up Lost Time', p. 97: 'History is meaning imposed on time by means of language: history imposes syntax on time'.

52 I am grateful to Gabrielle Spiegel for this suggestion.

53 E.M.C. van Houts, 'The *Gesta Normannorum Ducum*: a History without an End', *Proceedings of the Battle Conference on Anglo-Norman Studies* 3 (1980), pp. 106–18 and 215–20.

54 McKitterick, 'Constructing the Past', pp. 115–17. On the compilation of the Royal Frankish Annals see Roger Collins, 'The "Reviser" Revisited', in Alexander Callander Murray (ed.), *After Rome's Fall: Narrators and Sources of Early Medieval History* (Toronto, Buffalo and London, 1998), pp. 191–213.

55 Rosamond McKitterick, 'Political Ideology in Carolingian Historiography', in Yitzhak Hen and Matthew Innes (eds), *The Uses of the Past in the Early Middle Ages* (Cambridge, 2000), p. 167.

56 'Pippin was elected king according to the custom of the Franks, anointed by the hand of Archbishop Boniface of saintly memory, and raised to the kingship by the Franks in the city of Soissons. But Childerich, who was falsely called king, was tonsured and sent into a monastery', Royal Frankish Annals, *s.a.* 750, trans. Bernhard Walter Scholz (Ann Arbor, MI, 1970), p. 39. Rosamond McKitterick, 'Power in the Carolingian Annals', *English Historical Review* 115 (2000), pp. 16–18.

57 Royal Frankish Annals, *s.a.* 741, trans. Scholz, p. 37.

58 Paul E. Dutton, *The Politics of Dreaming in the Carolingian Empire* (Lincoln, NB, and London, 1994), pp. 86–7.

59 McKitterick, 'Constructing the Past', pp. 117 and 126–7.

60 Royal Frankish Annals, *s.a.* 780–1, trans. Scholz, pp. 58–9.

61 McKitterick, 'Constructing the Past', p. 126; 'Political Ideology', p. 167.

62 *Annals of St-Bertin*, *s.a.* 830, trans. Nelson, p. 21.

63 McKitterick, 'Political Ideology', pp. 167 and 173. It is also significant that these annals often circulated in codices with other historical texts: Collins, 'The "Reviser"', pp. 201–2 and, more generally, on the compilation of historical manuscripts see Rosamond McKitterick, 'The Audience for Latin Historiography in the Early Middle Ages: Text Transmission and Manuscript Dissemination', in Anton Scharer and Georg Scheibelrieter (eds), *Historiographie im frühen Mittelalter* (Vienna and Munich, 1994), pp. 96–114.

64 Consider, for example, the layout of the Parker Chronicle, CCCC, MS 173, from fo.4r. to foot of fo.16r.; discussed by Peter Clemoes, 'Language in Context: *Her* in the 890 *Anglo-Saxon Chronicle*', *Leeds Studies in English* n.s. 16 (1985), p. 31.

65 S. Foot, 'The Making of *Angelcynn*: English Identity before the Norman Conquest', *Transactions of the Royal Historical Society*, 6th series, 6 (1996), pp. 35–7; and 'Remembering, Forgetting and Inventing:

Attitudes to the Past in England after the First Viking Age', *Transactions of the Royal Historical Society*, 6th series, 9 (1999), pp. 197–200.

66 Kenneth Sisam, 'Anglo-Saxon Royal Genealogies', *Proceedings of the British Academy* 39 (1953), p. 298.

67 Craig Davis, 'Cultural Assimilation in the Anglo-Saxon Royal Genealogies', *Anglo-Saxon England* 21 (1992), p. 35. I shall revisit this issue at greater length in a future paper on perceptions of time in the Chronicle.

68 Compare Karl Löwith, *Meaning in History* (Chicago, 1949), p. 6.

69 Gabrielle Spiegel has discussed the ways in which medieval chroniclers used biblical typological interpretation, so that the past became an explanatory principle, a way of ordering and making intelligible the relationship between events separated by vast distance of time: *The Past as Text: The Theory and Practice of Medieval Historiography* (Baltimore, MD, 1997), pp. 92–7.

70 Bede, *De temporum ratione*, ch. 66, trans. Wallis, p. 158; Foot, 'The Making of *Angelcynn*', p. 32.

71 Augustine, *De civitate Dei*, eds Bernard Dombart and Alphonsus Kalb, *Aurelii Augustini Opera Pars XIV.1–2*, Corpus Christianorum, series Latina 47–8 (Turnhout, 1965), XVII.1, p. 551; this evocative translation is Marc Bloch's: *Feudal Society*, trans. L.A. Manyon, (London, 1961; 2 vol. edn, 1965), vol. 1, p. 90. See also Thomas Harrison, '"Prophesy in Reverse"? Herodotus and the Origins of History', in *Herodotus and his World: Essays from a Conference in Memory of George Forrest* (Oxford, 2003), p. 252.

72 Anglo-Saxon Chronicle, 937.

73 McKitterick, 'Political Ideology', p. 167.

74 Bede, *De temporum ratione*, ch. 67, trans. Wallis, pp. 239–40.

6

Functions of fiction in historical writing

Monika Otter

Despite its modern or even postmodern flavour, the notion that narrative history is a verbal construct, a textual artefact with its own poetics rather than a direct, uncomplicated reflection of events, would have come as no surprise to medieval writers and readers. It is really the opposite views that are modern. History as scholarly inquiry concerned with archival research and documentation is only about two centuries old, especially in the institutional academic form we know now. The mental habit of regarding historiography as a transparent medium with no literary substance of its own, a self-effacing text that simply shows things 'as they really were' ('wie es eigentlich gewesen'), derives from nineteenth-century historicism.[1] To classical, medieval and early modern Europeans, history was not a separate academic discipline, but a subsection of rhetoric (as was poetry and what we would call fictional narrative).

Not that they did not distinguish between these forms of textual production; on the contrary, medieval historians and literary theorists frequently defined *historia* as 'true' narrative, in opposition to the 'fabulae' of poets.[2] Medieval historiographers considered themselves beholden to truth, to what 'really happened'; Isidore of Seville went so far as to stipulate that all history should be based on eyewitness testimony.[3] This definition, obviously an inaccurate description of existing historiography and an impractical requirement for medieval writers, nonetheless underscores that medieval historiography makes as strong a truth claim as its later counterparts. The way this truth is established, however, may look different from ours. In the absence of eyewitness testimony, medieval and modern historians alike consult textual and documentary sources. Source criticism in the modern sense is not unknown to medieval writers, but it is less strongly developed and less easily performed, given the more limited access to information.[4] *Auctoritas*, the prestige and cultural acceptance of major texts, carries a greater weight for them than it does for us: whatever is reported in Bede, for instance, will be accepted unquestioningly by English historiographers of the high Middle Ages. So does collective memory and oral tradition, local or national. But there is also a kind of authority conveyed simply by the rhetorical and literary conventions of historical narrative. In the tradition of their Roman forebears, medieval historiographers think nothing of providing their historical characters with speeches – especially set pieces, such as a general addressing his troops before battle. The obvious modern question – How could the historian possibly know what was said on this occasion? – does not arise. Gaps in the historical record are often fudged or even filled in rather than acknowledged, the desire to create a continuous narrative outweighing the need for absolute fidelity to the documents. Characters and situations are often modelled on literary precedent; a historian would not hesitate, for example, to model his William the Conqueror on Julius Caesar, or to report a scene in terms strongly reminiscent of a biblical episode. In these instances, medieval historiography may seem less beholden to our standards of evidence than to a kind of rhetorical or textual 'truth' that resides in the literary form itself.

Most modern historians, scanning medieval sources for factual information rather than reading them for literary quality, simply get used to such narrative conventions and abstract from them. Until recently, it was unusual for historians to take literary approaches to historiography seriously; thus, they tended simply to disregard problematic elements, or consider them blemishes that detract from a text's source value. In some cases, editors even omitted portions of medieval chronicles they considered 'not historical'.[5] This attitude is becoming less common and many historians are becoming more receptive to what can be learned from literary studies and literary theory. Much historical information of a subtler kind – oblique commentary on the events reported, attitudes towards history, or meta-commentary on historiography – can reside in the narrative form.

Nonetheless, modern readers are often amazed to find plainly 'non-historical' elements incorporated in even the most sober, trustworthy medieval chronicles: miracles, 'mirabilia' (such as animals or landscape features with miraculous properties), mythical ancestors and foundation stories, even short narratives that have suspiciously close motif correspondences with folktales and romances. Even more problematic are those hard-to-classify narratives – Geoffrey of Monmouth's *History of the Kings of Britain* being the classic and widely influential example – that have all the trappings of history yet, to our minds, clearly belong in the realm of fiction. Exactly how medieval readers received such narratives is often not easy to ascertain. While there are indications that they did distinguish something like Geoffrey's capricious mythical history from more sober authorities such as Lucan, Eusebius or Bede, there are also plenty of indications that many readers took the story of Brutus, the mythical founder of Britain, and the story of Arthur, the ideal king and conqueror of the known world, as 'historical'.

To say anything of value about such hybrid and intermediary texts, besides simply dismissing medieval historians and their public as ill-informed and naïve, we need to sort out the terminologies, both modern and medieval. Even in modern terminology, everyday uses and popular notions often differ from scholarly practice and can confuse more systematic discussion. For instance, certain everyday usages to the contrary, the terminological pair 'history and fiction' is not synonymous with 'true and false'; indeed the two terms do not in any strict sense form a neat binary opposition at all. Our investigation is further complicated by incompatibilities between medieval and modern terminologies and categories. Medieval literary theory does not have a term comparable to our 'fiction' (let alone anything analogous to the vague and artificial bookstore category of 'non-fiction'). The terms that come closest ('fabula', 'argumentum') tend to be unhelpful. More importantly, medieval readers do not seem to have recognized fiction as a concept in any stable or explicit sort of way. At least one influential historian of literature, Hans-Robert Jauss, has indeed described the lack of that category as one of the chief markers of medieval 'alterity'.[6]

Yet, as Jauss's classic essay also taught us, the fascinating and vexing thing about studying medieval literature is that it seems both very 'other' and very familiar; our experience in reading it constantly oscillates between 'alterity' and 'modernity'. Medieval literary practice with regard to fiction may seem quite alien one moment, then, on closer reflection, not so very different from our own. Despite the incompatibility of terminologies, it seems that medieval readers, in practice, often did understand the concept of fictionality in much the same way we do. Perhaps this discrepancy between theoretical reflection and everyday practice should not surprise us; after all, the average modern reader is perfectly capable of reading, processing and enjoying fiction even though he or she is most likely unable to define the concept. Building on such an intuitive understanding of fictionality, medieval writers were able to play with fictionality as a matter of literary virtuosity, much as many modern writers do.

Genre distinctions were certainly permeable, probably more so than they are now. Saints' lives (supposedly the biographies of actual persons) may resemble secular romances and vice versa; romances or epics may be understood as historical 'fact'; historiography may freely borrow the narrative conventions of any or all of those genres. This permeability, on the other hand, may not be so very different from the modern situation after all. More than one commentator has observed that even now the distinction between (narrative) history and fiction is of interest precisely because they are not easily distinguishable by any formal means. Novels frequently give themselves the air of history; histories, especially those intended for a wider reading public, often borrow narrative techniques from the novel. A historical novel may be so carefully researched, may incorporate so much authentic source material, as to be very close to a 'real history'. How, then, do readers, modern or medieval, distinguish between the genres? One basic clarification – which may seem obvious once stated, but can nonetheless obviate a lot of terminological muddle – is that fictionality is not a function of truth *value* but of truth *claim*: not whether it corresponds to fact (however that may be ascertained), but how it asks to be taken by the reader.[7] Let us back up a bit to understand and justify this statement.

The one theoretical framework for understanding fictionality that the Middle Ages inherited from antiquity does indeed rest on truth values, and it is something of a blind alley. No Roman or medieval theorist, to my knowledge, was able to go any further with it than merely to repeat it, and we can dispose of it rather quickly. Classical rhetoric often distinguished 'historia' (true narrative) from 'fabula' (untrue narrative) and left it at that. This classification does not provide a satisfactory category for thoroughly canonical and admired but 'untrue' narrative masterpieces, such as the *Aeneid*, Ovid's *Metamorphoses* or Aesop's fables (let alone the less hallowed, vernacular productions of the Middle Ages, such as Arthurian romance); it merely lumps them together with lies. Recognizing this difficulty, many theorists hastened to defend certain kinds of *fabula* from the charge of lying, as long as they allegorically concealed, or dressed up, or sugarcoated, an acceptable moral or theological truth. This notion, often labelled the 'integumentum' ('garment') theory, still has some popular currency now, as anyone can easily test by asking random acquaintances to explain the concept of fiction. But it is plainly an afterthought, an ex post facto apologia for fiction; it is hardly the reason why fiction is written or read.

Other theorists, more interestingly, introduced a middle term between *historia* and *fabula*, often labelled 'argumentum'. Derived from forensic rhetoric, this term described hypothetical ('as if', 'let us suppose that . . .') narratives, which is indeed a promising way of theorizing fiction. *Argumentum* can function as a small and controlled space within 'serious' discourse (legal, homiletic, historical) in which narrative experimentation is possible.[8] But classical and medieval theorists do not do anything so interesting with the term. Having set up their basic framework in terms of truth value (historia = true, fabula = false), they quickly assimilate *argumentum* to that same framework and equate it (usually in the very next sentence after introducing it) with the 'verisimilar': that which 'resembles truth' and 'could have been true'. This notion, although of long standing (and still surprisingly popular, as, again, one can verify in a quick polling of one's friends), is plainly inadequate. Verisimilar fiction, such as a novel of manners that contains nothing implausible, is no more and no less 'true', and no more and no less 'fictional', than wildly fantastical narratives that bear no resemblance to reality as we know it.

Sheer truth value is clearly not of any help in defining fictionality, or in distinguishing history from fiction. This instantly excludes certain categories of question from our consideration. Elements in medieval historiography that are simply counter to 'fact' according to the criteria we use to establish it – misinformation, lack of adequate sources, outright lies, forgeries and tendentious reporting – are not in and of themselves usefully discussed in terms of fictionality (though they may sometimes be tangentially involved in questions that are pertinent). Nor should we attempt to consider here reports of phenomena that by our scientific standards are out of the realm of the rationally possible, but to medieval people may well have qualified, such as miracles performed by saints' relics or the monstrous hybrid animals sometimes reported in chronicles. These are cognitive dissonances (not of course limited to texts from the Middle Ages!) that we must simply live with.

If the medieval theorists' tendency to polarize history and *fabula* along a scale of truth values gets us nowhere, there is an almost equally unhelpful late twentieth-century tendency to treat history and fiction as essentially indistinguishable. The most prominent modern exponent of this line of thought is Hayden White. In his widely influential *Metahistory: The Historical Imagination in Nineteenth-Century Europe*, White reformulated for the late twentieth century the much older critical insight that narrative history is essentially a textual construct.[9] Medieval thinkers, as we have already seen, would likely concur. The standard medieval definition of history, derived from Isidore of Seville, 'Historia est narratio rei gestae, per quam ea quae in praeterito facta sunt dinoscuntur' ['history is a narration of events, through which that which occurred in the past is known'], seems to recognize at least a very basic distinction between the *res gesta* and its representation. History is not reality, not a sequence of events, but a means of 'knowing' past facts.

White is correct, of course, in saying that the nineteenth-century historicists' 'as it really was' (*wie es eigentlich gewesen*) begs the question of what the world 'really is', and how history could possibly go about representing it. No matter how scientific the historian's methods, no matter how meticulous and analytical the use of sources, no matter how dry the prose and how copious the tables and graphs, the resulting narrative will nonetheless select from all the various data that make up 'the world' or 'reality', and shape them in a way that reflects the historian's reasoned judgement of what is relevant and why, which events and phenomena can be seen as causally related, and so on. In other words, the historian's history is a narrative creation of his or her own making; and it is a text woven largely out of other texts, for most of the historian's raw materials, with the exception of the occasional archaeological data, are also textual. In his book (and in many important articles afterwards), White demonstrated that in ordering their narratives, or 'emplotting' their histories as he put it, historians follow essentially the same large-scale models or 'master tropes' used in fictional narrative.

Moreover, as White and others have shown, it is very difficult to describe in any philosophically satisfying way how a narrative can be 'true' to the world outside it, or even how exactly it *refers* to the world outside it. It is difficult enough to describe reference for individual terms: how words refer to concepts, and concepts to the world outside our heads, has been one of the main strands of western philosophical debate from Plato and Aristotle and the medieval 'universals' controversy up to the present. Matters become infinitely more complex when the terms are set in motion, as it were, and are woven together into propositions, not to mention the chain of propositions that makes up a narrative.[10] How can one historian's 'emplotment', or ordering of the large unstructured field of data that makes up reality, be said to be any more 'true' than another's? What *is* the referential relationship of a textual artefact composed now to a past state of affairs that no longer obtains and can no longer be accessed in any direct way at all? Modern thinkers are not the first to discover these fundamental difficulties; medieval writers often show a quite sophisticated awareness of them as well. As we shall see, these frustrating and intriguing uncertainties often provided an impetus for narrative experimentation, and a point of entry for fictional elements into historiography.

It is tempting to take the next step and to conclude that if history is textual in this sense, if the mechanisms of its outside reference are so hard to ascertain and describe, it is essentially no different from fiction. But simply equating textuality with fictionality is not particularly helpful, since in practice we do distinguish between historical and fictional narrative – and, barring a few borderline cases, are able to do so quite satisfactorily.[11] A pragmatic corrective to the textual perspective is therefore needed. Perhaps the most useful approach has been to view fictionality as a kind of 'contract' between author and reader, governing the reader's expectations as to the verifiability of the author's assertions. If you read a history of the French Revolution, you expect to be able to verify the existence, in surviving reality or in reliable traces, of places and characters; you expect to be able to find independent confirmation, in reliable sources, of the events described; you expect the dates to conform to the established calendar(s) and you expect that (unless otherwise explained) each event narrated be dated identically in other chronologies. If you discover serious and unexplained discrepancies, you legitimately feel betrayed by the author: he or she has violated the contract between you and you declare the history false. If you read Dickens' *A Tale of Two Cities*, on the other hand, or Büchner's play *Danton's Death*, you have no such expectations, even though they are recognizably set in revolutionary Paris and evoke familiar names and events. The spatial and temporal structure (the 'chronotope', as Mikhail Bakhtin would call it) of these fictions may well coincide in many particulars, or even entirely, with the historical record; some characters may coincide, by name and/or characteristics, with persons known to the historical record. But you will expect to encounter significant discrepancies and even entirely fictitious persons, places and sequences of events; and when you do, you have no legitimate cause for complaint. The contract between you and the author – made clear by sufficient signals in the text or in its packaging – does not *claim* that the narrative, its parameters and its events, be independently verifiable. Finally, in a less obviously historical fiction – such as Jane Austen's novels or Tolkien-style fantasies – you will, as a competent reader, know better than to go in search of any outside corroboration at all. A fiction is free to make up a world – coextensive with the text – with its own temporal and spatial structures, its own characters, its own boundaries, its own rules for plausibility, coherence and relevance; and these parameters may or may not resemble the everyday world we know. Fictional texts are beholden solely to these internal rules of their own.[12] Where historical fiction may, to some extent, employ a split reference, partly to its internal mechanisms and partly to a reality outside the text,[13] the reference of most fiction remains very largely within the text itself – its structures, its modes of emplotment, its chronotope, its characters.

Fiction, then, is not to be defined in terms of true or false at all; it is precisely the kind of narrative that brackets these questions, that makes outside reference and conventional truth values irrelevant.[14] Unless you are an unusually naïve or poorly acculturated reader, you know better than to ask if Gawain really existed.[15] There may be value in ascertaining the historicity of some elements in Arthurian romance, but that value lies in a different level of discourse altogether; for the purposes of the fiction itself, Gawain's 'reality' is simply beside the point. Modern readers do not need to understand the theory of fiction in order to respond to it quite competently. Likewise, even though there is no medieval word for fiction and the concept was not theorized in any cogent way, medieval readers and writers do seem to have understood fictionality essentially in much the same intuitive way that we do. When Wace says, famously, that he went to the legendary Arthurian forest of Broceliande and did not find a magical spring there ('I went there a fool, and returned a fool'),[16] he is not confusing fiction and history; rather, he is playing with his own, and his readers', quite fully developed sense of what fiction is.

Wace's example may remind us of an additional medieval wrinkle to the fiction/history distinction, namely the expectations conveyed by language (Latin versus vernacular) and, to a lesser extent, form (prose versus verse).[17] It has been argued that fictional playfulness is specifically and almost automatically a function of vernacular narrative. In this argument, Latin, connoting clerical and scholarly seriousness of purpose, not to mention divine sanction, in and of itself constitutes a contract of truthfulness and historical referentiality. Vernacular, being cut loose from the moorings of cultural seriousness, in and of itself constitutes something not unlike the fictional contract as we understand it, even if it undertakes to narrate history, as Wace does in the *Roman de Rou*.[18] But this distinction, as the example of Geoffrey of Monmouth shows, is at the very least too rigid. Latin does produce fiction, or fictional elements in historiography; and the vernaculars rapidly acquire genre markers of their own that convey the claim of historicity.

In what follows, I will briefly discuss two famous historiographical texts from twelfth-century England that make clever and meaningful use of fictionality: the *Gesta Regum Anglorum* by William of Malmesbury, the twelfth-century English historian most often complimented on his advanced source criticism and sound method; and Geoffrey of Monmouth's *Historia Regum Britanniae*, which can be read, among many other things, as an astute parody of William's brand of historiography.

Of all twelfth-century English historiographers, William of Malmesbury has earned the (relative) respect of modern historians; they consider him almost one of their own. He writes clear, elegant, no-nonsense Latin prose. For his two major historical works – the *Gesta Regum Anglorum* and the *Gesta Pontificum Anglorum* – he does extensive archival research, travelling all over England to view the documents preserved in various monasteries.[19] He practises source criticism, although his criteria may not always be the same as ours.[20] Rather than glossing over lacunae in his documentary record, he is quite frank about any lack of information or dearth of sources.[21] In the tradition of Roman historians, he does use literary devices such as dialogue and formal speeches, especially at moments of high drama;[22] and he does report the occasional 'mirabilia' that seem to us neither probable nor worthy of historical attention;[23] but apart from that, he is generally praised as an exceptionally trustworthy medieval witness.

Yet there are several puzzling sections in his *Gesta Regum*, which seem to stand out from the surrounding narrative in subject matter and style; William even frames them as digressions, often in almost apologetic tones.[24] These stories clearly demand to be read quite differently from the mainstream of William's historiographic narrative; in some cases he himself suggests allegorical or symbolic meanings for them; that is, he invites us to read them as metaphoric doublings of the events reported or interpretations of them. Yet many of the stories resist straightforward allegorizing; their function clearly is more complex. Perhaps most interesting among these is the sequence of stories in book 2, organized around the legend of Gerbert of Aurillac, the mathematician, philosopher, reputed necromancer and later Pope Silvester II.[25] In the version reported here, Gerbert, trained in the liberal arts and in magic by Saracens in Spain, is led to an underground treasure by a statue in Rome that is inscribed 'percute hic' ('strike here'). Many have tried doing so, but to no avail. Gerbert, in his ingenuity, is the only reader to interpret the message correctly: you must not strike the statue itself, but follow the shadow made by its outstretched finger at noon. Together with a servant, he digs at the spot so indicated and finds

a vast palace, gold walls, gold ceilings, everything gold; gold knights seemed to be passing the time with golden dice, and a king and queen, all of the precious metal, sitting at dinner, with their meat before them and servants in attendance; the dishes of great weight and price, in which work-manship outdid nature.[26]

Gerbert and his man cannot, however, touch anything in this underground realm, for it is uneasily poised, ready to disappear at the slightest disturbance: the bright light that illuminates it comes from a carbuncle, at which one of the golden figures is pointing a drawn bow and arrow. If anything is disturbed, the archer shoots the carbuncle and everything goes dark. This is precisely what happens: while Gerbert immediately grasps the situation and refrains from touching any-thing, his servant attempts to steal a golden knife. They barely manage to save themselves, having to leave the knife behind, of course.[27]

Next, William brings this story closer to home, and out of the books into first-hand oral testimony: 'I will repeat something I remember hearing in my boyhood from a monk of our convent who was a native of Aquitaine, a very old man and skilled as a physician'.[28] The story is given in direct speech, as told by the phys-ician himself. As a young man, travelling in Italy, this monk discovered 'a moun-tain with a hole in it, beyond which the local people thought that the treasures of Octavian had lain hidden since antiquity'.[29] Together with 12 companions, he ventures into the cave, which turns out to be a labyrinth strewn with the skel-etons of previous explorers. The young men, however, are better equipped than their hapless predecessors, for, using 'the device of Daedalus', they have brought a ball of thread to guide them back out. Finally, they arrive at a lake with a bridge; on the other side were 'wonderful great golden horses with riders too all of gold, and the other things in Gerbert's story'.[30] (The cross-reference to the Gerbert story is the more striking since the stories differ irreconcilably in many particu-lars, including the geographical location; the implication would seem to be that Gerbert's treasure can be found anywhere.) As in the previous tale, however, the golden world has mechanisms – almost comically mechanical ones – that protect it against intrusion: whenever one of the explorers sets foot on the end of the bridge, it tilts, setting in motion 'a bronze rustic with a bronze hammer with which he beat the water, filling the air with mist so that the sun and the sky were blotted out'. A second attempt to lift the treasure is also aborted, even though this time the young men are armed with 'the unutterable name of God', against which all magic is powerless. Finally, though, they bring a Jewish necromancer into the plan. He enters the mountain easily and brings up several of the objects the young men remember seeing beyond the lake, as well as some of the dust that turns everything it touches into gold – though, it turns out, only in appearance: 'it seemed so until it was washed in water; for nothing done by necromancy can deceive the eyes of a spectator when put into water'.[31] This assertion is backed up by another anecdote in which a young actor is turned, Apuleius-fashion, into an ass, and after a short career as a circus animal escapes by plunging himself into a lake.[32]

William does not entirely commit himself to the truth of these stories, though he adduces a number of arguments in their favour – among other things, the amazing story of Gerbert's terrible death, in which, in a desperate attempt to foil the devils waiting to pick up his soul, he has himself dismembered while still alive. Why, William asks, would a pope do such a thing unless there was some truth in the tales of his necromancy and pursuit of forbidden knowledge?[33] Since this story is equally fantastic as the ones it is introduced to support, we may not take it as strong evidence; and the overall effect of William's musings on the truth of the Gerbert stories is to mark them as belonging to a different realm of truth claim from the bulk of the history.[34]

Far from an irrelevant fantastic digression, I read this curious episode as one of the centres of the entire book. Gerbert and the Aquitanian monk are clearly author stand-ins, and the story of their exploits is what Lucien Dällenbach called a *mise en abyme*, a miniature replication of the story itself within the story.[35] The adventures of Gerbert, in other words, repeat the adventures of the historian author and mirror his epistemological situation. Massimo Oldoni argues convincingly that in Gerbert's frozen underground world, a world of automata, we have the historical facts *in potentia*, to be animated by the historian-magician's touch.[36] But despite the protagonists' considerable *ingenium*, the manoeuvre never appears to work, at least not in the narratives themselves. In one case, the material must not be touched at all; the whole underground world is poised to self-destruct, with possibly disastrous consequences for the intruder. In the other case, some material is brought to light, though by a character whose religion and methods mark him as suspect; but the objects disappear upon contact with water, as do, according to the narrator, all objects produced by magic. The obvious implication is that the material was not real to begin with: the demons guarding it are said to be up in arms not because they fear their property may be stolen, but because their 'inventions' might be 'refuted' (*refellere*).[37] Or else the process of lifting it from the cave somehow destroys its very reality. The magic performed by Gerbert is therefore dubious, not only in its morality (as all magic would be), but also in its efficacy. It is not clear that anything it can produce, or retrieve from its hiding place, has any substance or reality. History, the past, is seen as a separate place, literally 'reified', but dead – or ambiguously undead. It is untouchable, divided from the observer by various mechanical and magical barriers, and ultimately irretrievable.

Thus, William of Malmesbury, at the centre of his book, allows us to descend with him into his historiographical laboratory – and to watch him fail. The historian's position is acknowledged to be troublesome, even frightening. The very status of the world William describes, the referentiality of his writing and the possibility of truth-telling are put into play. Above all, we are invited to question how this historical narrative – or any historical narrative – can possibly 'bring to the surface', or reanimate, events of the past that are no longer with us; and what, then, it could possibly refer to. Even if the intent is not to shake our faith in his entire enterprise – we do continue to read the *Gesta Regum* as an informative and trustworthy work of history – William does voluntarily undermine the grounds for our trust. By inserting this fascinating episode into his history, and by carefully setting it off as a digression, William has opened up for himself a fictional space for reflecting on his art, for probing the limits of historiography, for all but suggesting that his historiography *is* fiction. But he stops just short of destroying its foundations completely; the episode is sufficiently contained to let the work proceed as 'history'.

William of Malmesbury is one of the writers Geoffrey of Monmouth addresses by name at the conclusion of his controversial *Historia Regum Britannie*. William of Malmesbury and Henry of Huntington, he says, can keep English history, as long as they stay away from the 'British' (that is, Welsh) history, which only Geoffrey knows. In other words, he warns them off his turf.[38] This playful (or perhaps aggressive?) challenge to his most eminent fellow historians has often been taken as one indication that Geoffrey's *Historia* is at least in part parodic, an explanation that has seemed attractive to many since it promises a way to deal with Geoffrey's clearly non-historical material. Whatever one's views on the 'historicity' of Arthur, he manifestly did *not* conquer Rome and most of the known world in the sixth century CE.[39] (It should be noted that parody does not necessarily mean a light-hearted or satirical spoof; it can be partly or entirely serious. In the strictest sense, parody simply means adopting the form of another text and filling it with new content. Satirical or humorous intentions are secondary.)[40] Other possible signals are conspicuous loose ends in the narrative; obscure details presented elliptically, as if they were common knowledge; oblique references and parallels to contemporary or near-contemporary events; the brazenness with which Geoffrey not only fills in the long gap in insular historiography but 'overstuffs' it, so that it conflicts with received chronology, including the unassailable authority, Bede.[41] But the strongest signal of all is the famous and most likely transparently spurious source fiction of the prologue. Geoffrey claims to derive all his information on the lost kings of Britain from 'a certain very ancient book in the British language . . . attractively composed to form a consecutive and orderly narrative', given to him by Archdeacon Walter and known or accessible to no one else but him.[42] The existence of such a book has never been demonstrated and is doubted by most. It is possible that Geoffrey had scattered written sources in Welsh and is rationalizing them with some poetic licence into the 'very ancient book'. He certainly had oral sources, as he says; and indeed, these oral traditions ('joyfully handed down . . . just as if they had been committed to writing') appear to congeal and metamorphose into the 'very ancient book' in the very next sentence. That is, Geoffrey presents his book as both an original composition and a faithful translation; as an oral tradition that is *just like* a book, and a book that, just like oral tradition, is not available for anyone's inspection or double-checking. He does and does not claim to be the author who gives shape to this narrative, and he does and does not claim the support of a textual *auctoritas*.

It is not Geoffrey's sincerity that is at issue here, but his skill in constructing a narrative that is curiously cut loose from all the moorings of referentiality, while preserving, at least in appearance, the same truth claim as any other history. It purports to refer truthfully to the sequence of events between the arrival of Brut and the decline of the Britons; but he has arranged it so that this reference is altogether unverifiable. The ancient book – the single source to which he claims, unverifiably, to be faithful – is the perfect embodiment of this attitude. So is the inserted text of Merlin's prophecies: an unstable centre of the work, almost incomprehensible in its obscure, apocalyptic images, which nonetheless claims to give meaning and interpretation to the surrounding narrative.

I do not intend to reopen here the question of Geoffrey's purpose in writing the *Historia*; the very fact that commentators have ranged so widely in their assessment of Geoffrey's politics is surely indicative of a purpose beyond simply taking sides in contemporary political struggles.[43] I am interested primarily in what can be ascertained about the reception of Geoffrey's work. What truth claim did medieval readers take the *Historia* to make? What sort of contract did they think Geoffrey was offering them? The *Historia* gained instant popularity, as the rich manuscript tradition and the many imitations, full-scale or of more limited scope, amply attest.[44] Some early readers called Geoffrey an outright liar – most famously, William of Newburgh, who attacks him for the impossibilities in his chronology, as well as the implausibility of Arthur conquering all those countries and nonetheless entirely escaping the notice of other historians.[45] Not all readers were so unforgiving, and even apparent criticisms may be less serious than they sound. Henry of Huntington, one of the earliest documented readers of Geoffrey's work, came across the *Historia* while visiting Robert of Torigni at Mont Saint-Michel. He hastily added an extended summary of it to his *Historia Anglorum*, but as a clear add-on, an inserted letter to one 'Warin Brito'.[46] That is, even though he says nothing to criticize or to question the truthfulness of Geoffrey's work, he treats it as a work apart from the mainstream, tremendously interesting but perhaps not quite on the same level of historicity, with not quite the same truth claim, as his other sources. Gerald of Wales does say that Geoffrey lies, but he nonetheless uses him as a source in his own account of the country and its history. And his famous joke about the mad prophet Meilyr, who could tell the truthfulness of a book by mere bodily contact with it, and had a horrendous fit whenever touched with Geoffrey's *Historia*, is a tongue-in-cheek quip directed as much at himself as at Geoffrey.[47] Gerald, that is, does seem to recognize that the *Historia*'s truth is at the very least not of the same kind as that of other historical sources; but he is not disturbed by this and enters into the spirit of Geoffrey's invention whenever it suits his purposes. In other words, he recognizes it as fiction.

Gerald and Henry are apparently not alone in this recognition. Geoffrey's source fiction seems not so much to have deceived readers as inspired imitations. Geoffrey became something of an *auctoritas* for those who engaged in creative history-making. Geoffrey's influence is often quite obvious; some texts mention him by name. The late twelfth-century *Historia monasterii de Abingdon*, for instance, uses the general outline of Bede's and Geoffrey's histories to insert the figure of the monastery's own legendary founder, Abennus, and a foundation myth concerning the finding of the 'Black Cross' (clearly modelled on St Helen's finding of the True Cross, which on Geoffrey's authority became associated with England).[48] William of St Albans, another twelfth-century monastic historian, even acknowledges Geoffrey's influence as he concocts a source fiction much like Geoffrey's ancient book.[49] In the mid-thirteenth century, St Albans' most famous historian, Matthew Paris, takes the book image one step further by having his monastic predecessors literally dig up an ancient book, vaguely described as 'in the old English or British language' (*antiquo Anglico, vel Britannico, idiomate*) in the nearby ruins of the Roman town of Verulamium.[50] A slightly later vernacular example is the richly inventive 'ancestral romance', *Fouke le Fitz Warin* (a kind of romanticized family chronicle or romance that served as family history), which creates an exuberant pastiche of all sorts of sources, but begins with a kind of parody and continuation of the story of Gogmagog in Geoffrey's *Historia* as a foundation myth for the Fitz Warin family.[51] Slighter echoes of Geoffrey can be found in other twelfth-century texts, such as the monastic foundation legends of Kirkstall and Selby.[52] Whatever the conscious intentions of any or all of these historiographers – whether or not they meant to deceive – Geoffrey seems to have inspired, or sanctioned, historical inventiveness in a way that Bede or William of Malmesbury did not. Whatever his contemporaries or near-contemporaries called this phenomenon in their own minds – fiction, lies or something else – it suggests that they had at least a semi-conscious sense that Geoffrey was not the same kind of source as Bede, a source perhaps more to be imitated and appropriated than quoted for evidence.

As these last examples show, fictionality – especially in the fluid genre landscape of the Middle Ages – can certainly be put to utilitarian uses, such as enhancing one's prestige or even defending property claims by giving oneself a genealogy or origin narrative. Curiously, overt fictionality is not always a disadvantage in such cases. St Edmund, the patron of the powerful abbey of Bury St Edmunds, for instance, in the twelfth century acquires a childhood narrative that is patently composed of folktale and romance elements, which, to us, would instantly disqualify it as a historical narrative; but that does not appear to have been its effect.[53] The same can be said of a similar monastic origin story, Matthew Paris's *Lives of the Two Offas*,[54] which is not only full of all sorts of folktale motifs (mute, blind and lame children suddenly cured; guilty or slandered queens set adrift in rudderless boats; letters maliciously exchanged; hermit confessors in the woods), but also doubles many of them so awkwardly in the lives of both Offa I and Offa II that the seventeenth-century editor, William Wats, was seriously angered by the patent implausibility of all this happening not once but twice! Yet the curious form of the 'prequel' permits a kind of suspension of reference, so that the artifice not only is not a problem but actually helps the writer's purpose. A prequel's truth claim is circular. Like its more straightforward cousin the sequel, it attaches itself to a narrative already established and accepted, but it simultaneously pretends to give rise and support to the already existing story. Even though it is patently an afterthought, it slips surreptitiously into an anterior position. We are to believe that the primary story came about because of what is narrated in the prequel; at the same time, we are to believe the prequel is true because we have already accepted that primary narrative. This closed system discourages outside reference – 'brackets' the question of any correspondence to outside events – and makes us rely on the logic of the text itself, its patterns, its repetitions, its question-answer, promise-fulfilment structures. Our sense of historical probability may rebel against the symmetry of the lives of the two Offas; our sense of narrative does not. Even if one does not fully 'believe' the story, it is easy to be seduced and convinced by its neat closure; and narratives of this kind have a way of establishing themselves as accepted truth, or quasi-truth, despite their patent implausibility.[55] Narratives like *The Two Offas*, St Edmund's childhood and countless other more or less fantastical foundation and origin legends, both secular and ecclesiastical, do not make the full truth claim of historiography; yet they serve almost all the same functions: providing ideological legitimization, collective identity, continuity and prestige.

Fictionality, then, enters medieval historiography quite easily, taking advantage of the fluid genre conventions. While the different truth claims of history and fictional narrative were certainly understood, crossovers and hybrids are possible and often useful. The functions of fiction vary widely, from ancestral legends for families, cities, monasteries or nations, with a quasi-truth claim that substituted for the real thing, to fictional elements in historiography that underscore the textuality and literariness of the narrative and playfully probe its truth claims. Each separate instance requires a careful reading, sensitive to its surroundings, its language, its literary techniques. In other words, it is important to take seriously the textuality of all historical narrative and read it with an eye to its literary structures as well as its documentary value.

Guide to further reading

Breisach, Ernst (ed.), *Classical Rhetoric and Medieval Historiography* (Kalamazoo, 1985).

Danto, Arthur, *Narration and Knowledge* (New York, 1985).

Haidu, Peter, 'Repetition: Modern Reflections on Medieval Aesthetics', *Modern Language Notes* 92 (1977), pp. 875–87.

Jauss, Hans Robert, 'The Communicative Role of the Fictive', in *Question and Answer: Forms of Dialogic Understanding*, ed. and trans. Michael Hayes (Minneapolis, MN, 1989), pp. 3–50.

Knapp, Fritz Peter and Niesner, Manuela (eds), *Historisches und fiktionales Erzählen im Mittelalter* (Berlin, 2002).

Mehtonen, Päivi, *Old Concepts and New Poetics: Historia, Argumentum, and Fabula in Twelfth- and Early-Thirteenth-Century Latin Poetics of Fiction* (Helsinki, 1996).

Mink, Louis O., 'Narrative Form as a Cognitive Instrument', in *The Writing of History: Literary Form and Historical Understanding*, eds Robert H. Canary and Henry Kozicki (Madison, WI, 1978), pp. 129–49.

Nykrog, Per, 'The Rise of Literary Fiction', in *Renaissance and Renewal in the Twelfth Century*, eds Robert D. Benson, Giles Constable and Carol Latham (Cambridge, MA, 1982), pp. 593–612.

Partner, Nancy F., *Serious Entertainments: The Writing of History in Twelfth-Century England* (Chicago, IL, 1977).

Partner, Nancy F., 'Making Up Lost Time: Writing on the Writing of History', *Speculum* 61 (1986), pp. 90–117.

Riffaterre, Michael, *Fictional Truth* (Baltimore, MD, 1990).

Spiegel, Gabrielle M., 'Genealogy: Form and Function in Medieval Historical Narrative', *History and Theory* 22 (1983), pp. 43–53, reprinted in *The Past as Text: The Theory and Practice of Medieval Historiography* (Baltimore, MD, 1997), pp. 99–110.

Spiegel, Gabrielle M., *Romancing the Past: The Rise of Vernacular Prose Historiography in Thirteenth-Century France* (Berkeley, CA, 1993).

Sternberg, Meir, *The Poetics of Biblical Narrative: Ideological Literature and the Drama of Reading* (Bloomington, IN, 1987).

Stierle, Karlheinz, 'Erfahrung und narrative Form: Bemerkungen zu ihrem Zusammenhang in Fiktion und Historiographie', in *Theorie und Erzählung in der Geschicte*, eds Jürgen Kocka and Thomas Nipperdey (Munich, 1979), pp. 85–119.

Stock, Brian, 'History, Literature, Textuality', in *Listening for the Text: On the Uses of the Past* (Baltimore, MD, 1990).

Trimpi, Wesley, 'The Quality of Fiction', *Traditio* 30 (1974), pp. 1–118.

Warning, Rainer, 'Staged Discourse: Remarks on the Pragmatics of Fiction', *Dispositio* 5.13–14 (1981), pp. 35–54.

Waswo, Richard, 'The History that Literature Makes', *New Literary History* 19 (1988), pp. 541–64.

White, Hayden, *Metahistory: The Historical Imagination in Nineteenth-Century Europe* (Baltimore, MD, 1973).

White, Hayden, 'The Value of Narrativity in the Representation of Reality', *Critical Inquiry* 7 (1980–81), pp. 5–27.

Notes

1   It has been pointed out that Ranke, who is usually cited as the author of the 'wie es eigentlich gewesen' dictum, did not mean it in quite such a naïve way; but, fairly or not, he has come to stand for a positivistic view that treats historiographical texts as more or less unproblematic sources of factual information and tends to disregard the poetics of historical narrative altogether, except in so far as it helps establish the trustworthiness of that source (see, for example, Stephen Bann, *The Clothing of Clio: A Study of the Representation of History in Nineteenth-Century Britain and France* (Cambridge, 1984), pp. 8–31). It should be noted that here and in the rest of the chapter, we are discussing narrative histories only – those, that is, which connect a series of events in a causally inflected sequence (what literary scholars would call a 'plot'); or, according to another minimal definition of narrative, those that describe 'a minimum of two states of affairs with a transition between them'. There are modern kinds of historiography that attempt to circumvent narrativity by describing synchronically a state of affairs at a single point in time, with statistical and other methods borrowed from the social sciences; but this form of historiography was not practised in the Middle Ages, or indeed any time before the mid-twentieth century.

2 See Peter Johanek, 'Die Wahrheit der mittelalterlichen Historiographen', in *Historisches und fiktionales Erzählen im Mittelalter*, eds Manuela Mesner and Fritz Peter Knapp (Berlin, 2002), pp. 9–25.

3 *Etymologiae* 1.41 See also Roger Ray, 'Bede's *Vera Lex Historiae*', *Speculum* 55 (1980), pp. 15–16.

4 See Nancy F. Partner, *Serious Entertainments: The Writing of History in Twelfth-Century England* (Chicago, IL, 1977), pp. 183–93, and Gabrielle M. Spiegel, 'Genealogy: Form and Function in Medieval Historical Narrative', in *The Past as Text: The Theory and Practice of Medieval Historiography* (Baltimore, MD, 1997), pp. 99–110 (reprinted from *History and Theory* 22 (1983), pp. 43–53).

5 Clearly stated, for instance, in *The History of the English by Henry, Archdeacon of Huntingdon,* ed. Thomas Arnold, Rolls Series 74 (London, 1879), pp. x–xxx.

6 Hans Robert Jauss, 'The Alterity and Modernity of Medieval Literature', *New Literary History* 10 (1979), p. 188, and Hans Robert Jauss, 'The Communicative Role of the Fictive', in *Question and Answer: Forms of Dialogic Understanding*, ed. and trans. Michael Hayes (Minneapolis, MN, 1989), pp. 4–10. On terminology, see also Benedikt Konrad Vollmann, 'Erlaubte Fiktionalität: Die Heiligenlegende', in *Historisches und fiktionales Erzählen im Mittelalter*, eds Manuela Mesner and Fritz Peter Knapp (Berlin, 2002), pp. 63–72.

7 For a wonderfully clear exposition of many of these issues, in regard to biblical narrative, see Meir Sternberg, *The Poetics of Biblical Narrative: Ideological Literature and the Drama of Reading* (Bloomington, IN, 1987), pp. 23–30.

8 Päivi Mehtonen has recently restated the case for *argumentum* as a fruitful and sophisticated theoretical category. *Old Concepts and New Poetics: Historia, Argumentum, and Fabula in Twelfth- and Early-Thirteenth-Century Latin Poetics of Fiction* (Helsinki, 1996). Peter von Moos sketches an intriguing theory of play and *argumentum* that strengthens the ties between it and most modern conceptions of fiction. *Geschichte als Topik: Das rhetorische Exemplum von der Antike zur Neuzeit und die 'historiae' im 'Policraticus' Johanns von Salisbury* (Hildesheim, 1988), pp. 276–85. Generally on *historia, fabula* and *argumentum*, see Mehtonen; Wesley Trimpi, 'The Quality of Fiction', *Traditio* 30 (1974), pp. 1–118; and Fritz Peter Knapp, 'Historische Wahrheit und poetische Lüge: Die Gattungen weltlicher Epik und ihre theoretische Rechtfertigung im Mittelalter', *Deutsche Vierteljahrsschrift* 54 (1980), pp. 581–635.

9 Baltimore, MD, 1973.

10 Not to mention the difficulty that the truly important referents of historiographical narrative are not things, people, or places but 'events'; and 'events' are arguably already an abstraction from reality – someone's

attempt to order and emplot raw data into a before/after, cause/effect. See Louis O. Mink, 'Narrative Form as a Cognitive Instrument', in *The Writing of History: Literary Form and Historical Understanding*, eds Robert H. Canary and Henry Kozicki (Madison, WI, 1978), pp. 129–49.

11   Borderline cases for medieval readers include the ones mentioned above, especially Geoffrey of Monmouth. Modern borderline cases, it should be noted, always arise where the truth *claim* or genre seems ambiguous. Had Oliver Stone clearly presented *JFK* as fiction, had Edmund Morris's hybrid text about Ronald Reagan not claimed to be a biography, they might not have been as controversial.

12   See John R. Searle, 'The Logical Status of Fictional Discourse', in *Expression and Meaning: Studies in the Theory of Speech Acts* (Cambridge, 1979), pp. 58–75 (reprinted from *NLH* 6 (1974–75), pp. 319–32) and Karlheinz Stierle, 'Erfahrung und narrative Form: Bemerkungen zu ihrem Zusammenhang in Fiktion und Historiographie'; in *Theorie und Erzählung in der Geschichte*, eds. Jürgen Kocka and Thomas Nipperdey (Munich, 1979), pp. 85–119.

13   See Andreas Wetzel, 'Reconstructing Carthage: Archaeology and Historical Novel', *Mosaic* 21 (1989), pp. 13–23.

14   For classic statements of this definition with particular reference to medieval literature, see Peter Haidu, 'Repetition: Modern Reflections on Medieval Aesthetics', *Modern Language Notes* 92 (1977), pp. 875–8, and Michel Zink, 'Une mutation de la conscience littéraire: le langage romanesque à travers des exemples français du XIIe siècle', *Cahiers de Civilisation Médiévale (xe–xiie siècles)* 24 (1981), pp. 3–27.

15   Those people who do ask that question are either asking it in a form that goes well beyond the historical narrative – that is, they are not checking up on the truth of Chretien de Troyes' romances, but searching for an extraliteral character that quite remotely served as a 'model' for the literary character. Or else they are playing an elaborate fictional game of their own, and they are usually more or less aware of it.

16   Wace, *Le Roman de Rou*, ed. A.J. Holden, Société des Anciens Textes Français (3 vols, Paris, 1970–73), vol. 2, pp. 121–2, lines 6373–98.

17   Haidu, 'Repetition' and Gabrielle M. Spiegel, *Romancing the Past: The Rise of Vernacular Prose Historiography in Thirteenth-Century France* (Berkeley, CA, 1993).

18   Several critics have made this case, but see especially Haidu, 'Repetition'.

19   Rodney M. Thomson, *William of Malmesbury* (Woodbridge, 1987), pp. 1–38, and Gransden Antonia Gransden, *Historical Writing in England* c.*550 to* c.*1307* (Ithaca, NY, 1974), pp. 166–85.

20   Spiegel, 'Genealogy', p. 102, characterizes his method as 'a kind of primitive sic et non'; see also Thomson, *William of Malmesbury*, pp. 16–18.

21 This, though, could also be considered a literary device. William sees himself as restarting a tradition of English historiography after a long hiatus, reconnecting directly with Bede. Hence, his complaints about a lack of records – due in part to the Norman Conquest and in part to the negligence of earlier generations – helps him construct a very self-conscious writerly persona. In other words, what appears to us to be 'modern' about William of Malmesbury's method may in many cases have quite different objectives than we might assume if we extrapolate from modern practice.

22 For instance, the deathbed speech of Pope Gregory VI, defending himself against false accusations (William of Malmesbury, *Gesta Regum Anglorum: The History of the English Kings*, ed. and trans. R.A.B. Mynors, completed by R.M. Thomson and M. Winterbottom (2 vols, Oxford, 1998–99), vol. 1, pp. 368–77). In another instance, reporting in full a long crusading sermon, William acknowledges that he is composing the speech himself, 'preserving intact the sense of what was said; the eloquence and force of the original who can reproduce? We shall be fortunate if, treading an adjacent path, we return by a circuitous route to its meaning' (vol. 1, pp. 598–9). See Marie Schütt, 'The Literary Form of William of Malmesbury's "Gesta Regum"', *English Historical Review* 46 (1931), pp. 255–60, and Joan G. Haahr, 'William of Malmesbury's Roman Models', in *The Classics in the Middle Ages*, eds A. Bernardo and S. Levin (Binghamton, NY, 1990), pp. 165–73.

23 For instance, the story of the 'witch of Berkeley' (vol. 1, pp. 376–81) and the series of sinister events reported for the thirteenth year of William Rufus's reign (vol. 1, pp. 570–1), explained as portents of his violent death.

24 For example, vol. 1, pp. 278 and 294.

25 The story is told, in sometimes quite different versions, in a number of sources. See Massimo Oldoni, 'Gerberto e la sua storia', *Studi Medievali* 3rd series, 18 (1977), pp. 629–704 and '"A fantasia dicitur fantasma." Gerberto e la sua storia, II', *Studi Medievali* 3rd series, 21 (1980), pp. 493–622, and 24 (1983), pp. 169–245. For a very convincing alternative reading of the episode, see David Rollo, *Glamorous Sorcery: Magic and Literacy in the High Middle Ages* (Minneapolis, MN, 2000), pp. 3–23.

26 Vol. 1, pp. 285–7.

27 Vol. 1, pp. 286–7.

28 Vol. 1, pp. 288–9.

29 Vol. 1, pp. 288–9.

30 Vol. 1, pp. 288–9.

31 Vol. 1, pp. 290–1.

32 Vol. 1, pp. 292–3.

33 Vol. 1, pp. 282–3.

34 Generating discussions about truth and probability would seem to be one of the functions of the Gerbert myth; most medieval narratives about him are framed by similar doubts. See Monika Otter, *Inventiones: Fiction and Referentiality in Twelfth-Century English Historical Writing* (Chapel Hill, 1996), pp. 110 and 189–90 (n. 60).

35 Dällenbach, *The Mirror in the Text*, trans. Jeremy Whiteley with Emma Hughes (Cambridge, 1989). Dällenbach gives a highly suggestive list of common author stand-ins, 'qualified personnel from among those who specialize in, or make their living from, the truth', such as scientists, clergymen, librarians – but also madmen, drunkards, and dreamers (p. 53).

36 Oldoni 2, pp. 557–9.

37 Or, as Thomson translates, they 'were no doubt jealous of the name of the Lord which could make nonsense of all their inventions' (vol. 1, p. 291).

38 *The Historia Regum Britannie of Geoffrey of Monmouth, I: Bern, Burgerbibliothek, MS. 568*, ed. Neil Wright (Cambridge, 1985), pp. 146–7.

39 Robert W. Hanning, *The Vision of History in Early Britain* (New York, 1966), pp. 135–6; Valerie I.J. Flint, 'The *Historia Regum Britanniae* of Geoffrey of Monmouth: Parody and Its Purpose', *Speculum* 22 (1979), pp. 447–68.

40 On 'serious' parody, see Linda Hutcheon, *A Theory of Parody: The Teachings of Twentieth-Century Art Forms* (New York, 1985), pp. 50–68.

41 R. William Leckie, Jr, *The Passage of Dominion: Geoffrey of Monmouth and the Periodization of Insular History in the Twelfth Century* (Toronto, 1981), pp. 25, 66–7, 95–97.

42 'Quendam Britannici sermonis librum uetustissiumum', *Historia Regum*, ed. Wright, p. 1; in the convenient and readable, though occasionally unreliable, translation by Lewis Thorpe, Geoffrey of Monmouth, *The History of the Kings of Britain* (Harmondsworth, 1983), p. 51.

43 The main point of contention is whether Geoffrey, himself apparently a Breton raised in Wales, is championing the Welsh (or a pan-Celtic resurgence) over the Anglo-Norman ruling class. This position has recently been rearticulated with more convincing arguments than before: John Gillingham, 'The Context and Purposes of Geoffrey of Monmouth's *History of the Kings of Britain*', *Anglo-Norman Studies* 13 (1991), pp. 99–119. Several recent commentators have used the insights and terminology of postcolonial theory to describe Geoffrey's position as a writer from a 'subaltern' colonized population. While serious questions have been raised about the applicability of postcolonial paradigms to the Middle Ages, in this particular question the approach has been fruitful: postcolonial theory helps us account for *both* Geoffrey's trickster-like

pro-British sentiments and his 'collaboration' with the Anglo-Norman elites, and also his apparent finessing in playing various sides in the royal succession wars between Stephen and Mathilda. See especially Michelle R. Warren, *History on the Edge: Excalibur and the Borders of Britain, 1100–1300* (Minneapolis, MN, 2000), pp. 25–60.

44 See Julia C. Crick, *The Historia Regum Britannie of Geoffrey of Monmouth, IV: Dissemination and Reception in the Later Middle Ages* (Cambridge, 1991), and Leckie, *The Passage of Dominion*.

45 William of Newburgh, *The History of English Affairs, Book I*, eds and trans. P.G. Walsh and M.J. Kennedy (Warminster, Wiltshire, 1988), pp. 32–5.

46 Henry, Archdeacon of Huntingdon, *Historia Anglorum: The History of the English People*, ed. and trans. Diana Greenway (Oxford, 1996), pp. 558–63. In a note, Greenway suggests the possibility that Warin, whose identity is unknown, might be a fictitious person, perhaps a generic 'Brito' (Breton, or even 'Briton', i.e. Welsh), to whom Geoffrey's *Historia* ought to be of particular interest (p. 559, n. 2).

47 See Otter, *Inventiones*, pp. 152–4.

48 *Chronicon Monasterii de Abingdon*, ed. Joseph Stevenson, Rolls Series 2, 2 vols (London, 1858), vol. 1, pp. 6–7.

49 'Alia Acta SS Albani, Amphibali et sociorum . . .', *Acta Sanctorum* (Paris, 1867), pp. 129–38.

50 Matthew Paris, 'Gesta Abbatum Monasterio Sancti Albani', in *Gesta Abbatum Sancti Albani*, ed. Henry Thomas Riley, vol. 1, Rolls Series 28.4 (London, 1867), pp. 1–324, reprinted n. p., Kraus Reprint, 1965.

51 *Fouke le Fitz Waryn*, eds E.J. Hathaway *et al.* (Oxford, 1975). On English ancestral romance, see M. Dominica Legge, *Anglo-Norman and its Backgrounds* (Oxford, 1963), pp. 139–75, and Susan Crane, *Insular Romance: Politics, Faith, and Culture in Anglo-Norman and Middle English Literature* (Berkeley, CA, 1986), pp. 53–91; for vernacular family histories in France, see Spiegel, 'Genealogies' and *Romancing the Past*.

52 *The Foundation of Kirkstall Abbey*, ed. and trans. E.K. Clark, vol. 2 (Leeds, 1893), and 'Historia Selebeiensis Monasterii, quod fundatum est in Anglia . . .', *The Coucher Book of Selby*, ed. J.T. Fowler (Durham, 1890), pp. 1–54.

53 Galfridus de Fontibus, 'De Infantia sancti Eadmundi', in *Memorials of St. Edmund's Abbey*, ed. Thomas Arnold, Rolls Series 96 (London, 1890), vol. 1, pp. 93–103. In Denis Piramus's late twelfth-century Life of St Edmund (Denis Piramus, *La Vie Seint Edmund Le Rei*, ed. Hilding Kjellman (Gothenburg, 1935), this childhood narrative is seamlessly fused with the existing legend of his adult life and martyrdom and thus becomes permanently part of his story.

54 'Vitae Duorum Offarum', ed. William Wats (London, 1639); a large portion of the text is also printed in R.W. Chambers, *Beowulf: An*

Introduction to the Study of the Poem with a Discussion of the Stories of Offa and Finn (Cambridge, 1921), pp. 217–43. See Monika Otter, 'La Vie des deux Offa, L'enfance de Saint Edmond, et la logique des antécédants', *Médiévales* 38 (2000), pp. 17–34.

55 For other, more famous examples of such fabulous ancestors, see Jacques Le Goff, 'Melusina: Mother and Pioneer', in *Time Work, and Culture in the Middle Ages* (Chicago, IL, 1980), pp. 205–22; Donald Maddox and Sara Sturm-Maddox (eds), *Melusine of Lusignan: Founding Fiction in Late Medieval France* (Athens, GA, 1996); and Thomas Cramer, *Lohengrin: Edition und Untersuchungen* (Munich, 1971), pp. 68–123.

Part 3

Historicizing sex and gender

7

Historicizing sex, sexualizing history

Jacqueline Murray

For over 25 years historians have been exploring the meaning and experience of sex in past societies. Two works, in particular, served to focus scholarly attention on this area. These two works, *Christianity, Social Tolerance, and Homosexuality* by John Boswell[1] and *The History of Sexuality* by Michel Foucault,[2] have inspired much of the research into the history of sex. Innovative, controversial and very different in approach, they also signalled many of the tensions and ambiguities that continue to characterize the study of the history of sex. Boswell looked into the past to find the historical roots of the gay sexuality he saw in the modern world. Foucault argued that modern values about sexuality were foreign to the people of the past. These differing approaches highlight the tensions that exist between history as the study of change over time and sex as stable and common across time, place and culture. Some of the most challenging and insightful research into the Middle Ages can be found at the interstices of change and continuity, belief and act, body and soul. The tensions, then, have been creative ones that continue to impel historians to search for ways to understand the experience of human sexuality in the past.

Sex is a concept fraught with ambiguity and obscured by multiple layers of meaning. It is particularly vulnerable to anachronism, value-laden interpretations and, perhaps much worse, to preconceptions and assumptions. As understood in biology and physiology, sex refers to biological difference, male and female, that is, to women and men's differently sexed bodies. Sex, however, encompasses much more than biological difference and the reproduction of species. Even as a noun, sex is an active word indicating the desire for or engagement in some form of sexual activity. This might most frequently be understood to refer to sexual intercourse between a man and a woman, but that hardly exhausts its protean meanings. 'Having sex' can refer to a diversity of sex acts between/among various combinations of individuals, or to solitary activities. But sex is also far more complicated than bodily difference or bodily acts. It also occupies a psycho-emotional realm, linked to the fundamental core identity of an individual. In the seventeenth century, John Donne was one of the first writers to express this integrated meaning of sex in his poem, 'The Primrose'.[3] This articulation of sex has become increasingly central to our understanding of the whole human being. In this respect, sex embraces emotions, desires, eroticisms, preferences, identities and orientations. And, as with so many aspects of human experience, the matrix of desire and pleasure is partly culturally constructed and partly psychologically determined, and always profoundly individual. In all these multiple layers of meaning, sex has a history as complex and curious as the Middle Ages itself.

Nature, sex difference and the medieval body

The body, as a phenomenon of nature, generally appears stable and immutable, except over evolutionary time. In this view, bodies would necessarily be entities without a history. But even at its most basic and seemingly fixed aspect, even at the level of the biological difference between the sexes, we find that the human body was understood differently in the medieval period, with the result that society regulated the body according to contemporary understandings, and men and women experienced their sexed bodies differently than is the case today.

Although the philosophical and scientific arguments known to medieval scholars, which could be traced back to Plato and Aristotle, and the Christian theological interpretations that developed in Roman times and later, were the preserve of the educated, primarily clerical, elite of society, they nevertheless provided a firm theoretical foundation for understanding the natural order and justifying the hierarchy of the sexes. Medieval society inherited from the ancients the belief that men and women essentially had the same bodies. Sex difference was the result of different complexions or humours that caused men's genitals to be external while women's were inverted and inside the body. The organs themselves, however, were homologous, so that the ovaries were viewed as internal testicles and the vagina appeared as an inverted penis. This so-called 'one sex' understanding of sex difference[4] had many implications for how human sex difference and sexual functioning were understood. The balance of the humours and the complementary nature of male and female physiology implied that sex difference was not something that was fixed, stable or uniform. In humoral medical theory, men were considered to be hotter and drier, while women were colder and moister. This was, however, only a difference of degree, with the result that, as a result of illness or external causes or the inborn nature of an individual, a woman might be hotter or a man colder than the norm. These characteristics could influence an individual's sexual nature, which could come to resemble that of the other sex. Female social subordination mirrored the natural order, in which the female was considered an imperfect and less-developed male. Thus, in the Middle Ages, science and theology reinforced each other, affirming and justifying the dominance of the male and the subordination of the female both in the natural order and in the order of salvation.

The humours were believed to influence how a man experienced sexual arousal and spontaneous genital responses such as seminal emissions. For example, a hot nature could result in a man experiencing 'the fires of lust'. In the case of the hermit Godric, 'his flesh frequently burned with such ardent desire that many times the seminal fluid was seen discharging itself through natural channels'. To 'cool' his sexual desire and control his emissions of semen, Godric, like dozens of other holy men before and after him, immersed himself in icy water. Conversely, Robert Grosseteste, bishop of Lincoln in the thirteenth century, suggested that the reason he did not experience seminal emissions was not because he could control his lust, but because he had a colder nature than other men.[5] Perhaps more interesting for how it reveals that belief in a different physiology could influence how a person experienced her own body, is Heloise's assertion that she would burn with such sexual desire for the absent Peter Abelard that she, too, experienced 'movements of the flesh': 'Even in sleep I know no respite. Sometimes my thoughts are betrayed in a movement of my body'.[6] Heloise employed the same language used by male authors to describe nocturnal emissions to explain her own experience of erotic dreams and orgasm. This followed logically from the prevalent belief that men and women's bodies were essentially the same and so would behave analogously. Thus, Heloise, an educated woman with extensive knowledge of Latin literature, experienced her body in a way very different from modern women, in part because of prevailing beliefs and conceptual vocabulary. It was both logical and necessary for her to appropriate the masculine language of nocturnal emissions to describe her experience of erotic dreams and orgasm. This was the only language available to her to describe her experiences and it was a language that would have been deemed suitable given the homologous nature of male and female bodies and their associated sexual desires and experiences. Thus, beliefs about the body and sexuality significantly informed Heloise's experience of her own body.

How medieval bodies, so differently constructed in cultural understanding from those of our contemporary world, could lead to very different understandings about the nature of men and women and appropriate gender behaviour is illustrated in a thirteenth-century fabliau entitled *The Lady Who Was Castrated*.[7] In this story, a domineering woman is conquered by her son-in-law, who will not tolerate his mother-in-law's refusal to submit to her husband specifically and to patriarchal authority in general. The son-in-law asserts that she has 'balls like ours, and that's why your heart is so proud'. He has a bull castrated and the steaming bloody testicles brought to him. He then seizes his mother-in-law, holds her so she cannot see what he is doing, cuts into her thigh, and pretends to castrate her. Afterwards, he threatens his wife with similar treatment: '[I]f I see that you want to rebel against me, your balls will be removed, just as we have done to your mother. For it is just such testicles that make women proud and foolish'.

This unusual tale works on many levels of meaning and is both complicated and confusing. It demonstrates the perceived link between the sexed body and appropriate gender behaviour. The metaphorical 'castration' of an assertive, even manly, woman serves to reinscribe appropriate gender roles and their link to physical characteristics. The 'castration' was understood to 'readjust' the woman's humours towards the feminine end of the continuum, although the narrator makes clear that the castration is fictional, an elaborate trick. Nevertheless, the castration also works on another level of emotional drama as, in the story's progression, the bull's testicles are accepted as the symbolic source of the woman's dominance both by the son-in-law and by the woman herself. Both the characters seem to forget that the woman has been dominant without male organs. In this way, the bull's testicles function narratively as a physical metaphor for masculine dominance. The symbolic castration of the woman has real narrative consequences and returns her to the appropriate female social behaviour. Such a view of sex difference may account in part for why medieval society enforced such rigid gender regulations. Actual sex slippage was theoretically possible. Humours grew hotter or colder, moister or drier, and accordingly one's place on the sex continuum could move further away from one sex and closer to the other. Individuals, as well as society as a whole, had a vested interest in developing ways in which to clarify sex and gender and secure each person's appropriate place on the continuum.

The physiological imperatives of the sexed body could also contravene the moral code of the Catholic Church. One of the many ways that a healthy balance of the humours could be maintained was through the judicious expulsion of pent-up bodily fluids. Medical treatises might advise sweating, urination, elimination or bloodletting to treat diseases. More problematic for moralists, was the advice for the treatment of 'retention of the menses'. A woman who was unable to menstruate would suffer a variety of maladies that resulted from her inability to balance her humours naturally. Physicians continued to provide remedies for this malady, remedies which in antiquity had been used as abortifacients. Thus, under the more innocuous guise of treatment to expel retained menstrual blood, advice on abortion continued to circulate across medieval Christendom.

Men, on the other hand, might suffer from a build-up of semen that needed to be expelled to restore health. This medieval lore led to various stories about men whose faith was tested by the temptation to expel semen in a sinful manner. In three such stories from the twelfth century, with quite different endings, the man was given the choice of sinning or dying. According to Gerald of Wales, King Louis VII of France was told he would die unless he had intercourse.

Unfortunately, the queen was absent from court, so his physicians and courtiers advised him to have sexual intercourse with another woman in order to be cured. The king refused to commit adultery and lose his soul in order to save his life. God rewarded his virtue by curing him anyway.[8] An equally miraculous cure saved Thomas, archbishop of York, from breaking his vow of chastity, despite the urgings of both his physicians and clerical advisors that it was better for him to have intercourse with a prostitute than to die.[9] A similar situation occurred at Louvain, where an archdeacon was elected to the episcopacy, despite his protestations that he could not remain celibate and his inevitable sin would be exacerbated if he were in a higher order of the clergy. Gerald of Wales reports that within a month of his elevation, the man's penis swelled to an immeasurable size and when his friends and family realized how ill he was, they urged him to have sex. The unfortunate man, however, lamenting that he would have preferred to have been a living archdeacon than a dead bishop, refused the advice. Although his situation grew worse, he nevertheless did not yield to the remedy of sexual intercourse, in his view a sinful temptation, and died shortly afterwards. Gerald of Wales concluded that this heroic exercise of continence earned the man a place among the saints and martyrs.[10]

While semen and sex itself could be perceived as powerful and therapeutic, there was also a certain fragility and vulnerability associated with the expulsion of semen: excessive ejaculation could weaken a man to the point of death. In *The Art of Courtly Love*, Andreas Capellanus noted that men who were sexually active were physically weaker and aged more quickly than their chaste brothers.[11] More ominously, John of Salisbury recounted the case of Ralph, count of Vermandois. Ralph had married three times, a number which medieval churchmen would have judged to be excessive, even inappropriate. While he was recovering from an illness, Ralph's physicians advised him to abstain from sexual intercourse. According to John, however, Ralph 'disregarded this warning, for he was very uxorious. When the doctor detected from his urine that he had done so, he advised him to set his house in order, as he would be dead within three days'.[12] John concludes with the moral that 'this man perished through the vice by which he was most passionately enslaved, for he was always dominated by lust'. In this example, the expulsion of semen in sexual intercourse, even within marriage and with one's legitimate spouse, was fraught with both physical and spiritual danger.

Regulating the sexual body

The regulation of sexual behaviour by laws and by customs is an aspect of sex that is easily visible, recognizable and familiar, and has been subject to traditional modes of historical analysis. This is one area of sex that has had a clear history: the changes over time in the social control of sexual behaviour. All societies develop norms to govern who may engage in legitimate sexual relations. Most societies, for example, forbid sexual intercourse between parents and children or siblings. Medieval Christian Europe extended these widespread sexual taboos much further and forbade intercourse between relatives in the seventh degree of blood relationship (consanguinity). Moreover, sexual relations between people related by marriage (affinity) or spiritual ties (godparenthood) were also forbidden.

Sexual taboos had a tremendous influence on political, economic and social relations, since they defined and delimited legitimate marriage. Yet, while they may appear to have been primarily a form of social regulation, at heart these taboos were fundamentally sexual. For example, in the eighth century, Boniface argued that the impediment of affinity was actually created not by marriage but rather by the act of sexual intercourse, whether it were marital intercourse, adultery or fornication. Consequently, canonists, theologians and pastors developed mechanisms to deal with the implications of prohibited affinities which arose in various extramarital contexts. A single act of sexual intercourse with a prostitute established a permanent impediment of affinity among her clients and their relatives. Troubling questions could arise, such as: Was a prostitute responsible for ensuring that she did not accept two brothers as clients, a situation with incestuous implications? A more complicated extension of this logic forbade a man to marry the sister of any of the other clients of a prostitute whom he had frequented. On another level, what happened if a man married the sister of a woman his father had raped? These kinds of conundrums may sound suspiciously like scholastic riddles, but canon lawyers took seriously the spectre of unconscious incest, illegal marriages and people unaware that they were living in a state of mortal sin. From the thirteenth century onwards, these were matters priests were supposed to explore with their lay parishioners in the sacrament of confession.

The sexually created and historically contingent nature of affinity is highlighted further by the fact that anal intercourse (either between a man and a woman or between two men) was not considered to create the bond of affinity. In other words, a barrier of incest did not result from *any* sexual activity, but rather a specific act: vaginal intercourse between a man and a woman. The important distinguishing act that created the impediment was the penis penetrating the woman and ejaculating semen in her vagina. Anal intercourse was not considered to create the impediment of affinity because it did not result in the 'mingling of seed'. Thus, as early as the mid-ninth century, Hincmar of Reims dismissed allegations by Lothair, King of Lotharingia, that his wife, Teutberga, was a technical virgin but had committed incest with her brother, by anal intercourse, had conceived a child in this manner and had subsequently procured an abortion.

The astonishing array of allegations levelled by Lothair reveals a number of complexities pertaining to the medieval understanding of sex difference. The 'one-sex' theory meant that it was believed that both men and women could, and indeed must, 'ejaculate' semen in order for conception to occur. Contemporary historians have argued that sexual pleasure must have been considered to be equally important for medieval men and women since both contributed equally to conception. This is an assumption, however, that needs to be interrogated carefully.

The medical belief in the two-seed theory of conception has sometimes been seen as complementary to the theological doctrine of the conjugal debt. This concept was based on Paul's injunction that:

> The husband should give to his wife her conjugal rights, and likewise the wife to her husband. For the wife does not rule over her own body, but the husband does; likewise the husband does not rule over his own body, but the wife does. Do not refuse one another except perhaps by agreement for a season, that you may devote yourselves to prayer; but then come together again, lest Satan tempt you through lack of self-control (1 Corinthians 7:3–6).

The conjugal debt, then, was an obligation between married people to give each other sexual access freely, whenever and wherever it was requested. Only if the couple mutually agreed to sexual abstinence was it possible to deny sex to one's spouse without incurring serious sin. Paul wrote to new Christian communities, in an atmosphere of religious enthusiasm, when a convert to Christianity might embrace chastity as a spiritual discipline and refuse to have marital intercourse with a spouse who was unwilling to live in chastity. Thus, the unilateral actions of one spouse could disrupt the marital relationship and expose the other to the more serious temptations of adultery, divorce, masturbation or even same-sex sex acts. As time went on, however, layers of interpretation and changes in social context modified the practical understanding of the conjugal debt. Rather than a protection of marital harmony against unilateral, selfish chastity, the conjugal debt came to reflect an understanding of sexual desire as virtually uncontrollable lust. In order to avoid the greater sins of adultery, prostitution, masturbation or sodomy, spouses were required to grant each other sex 'whenever and wherever it is asked'. If a spouse wanted sex at a prohibited time, during Lent for example, or while the woman was menstruating, or in a prohibited place such as a cemetery or church, the spouse who demanded the debt sinned seriously, but the one who acceded to it was considered to be free from any guilt associated with the inappropriate sex act.

Because the conjugal debt stressed reciprocity, some scholars have suggested it afforded women sexual equality within marriage. Canon law gave women an equal right to demand sexual satisfaction, as long as it was within legitimate marriage. An alternative interpretation, however, is that, given prevailing social views about the passive and subordinate place of women within marriage, genuine equality in the area of sex was highly unlikely in practice.[13] Indeed, the asymmetry of the conjugal debt was acknowledged by pastors and moralists such as Thomas Aquinas, who observed that: 'The husband is bound to render [the conjugal debt] to his wife when she does not ask'. A husband was supposed to watch for signs that his wife desired sex but was too embarrassed or ashamed to ask in a forthright fashion. Other writers, however, seemed to recognize that a wife's right to sexual reciprocity was limited. For example, according to some moralists, if a woman's husband had already rendered the debt and was tired and physically depleted, she did not have the right to ask for sex again. If she did, she was considered to be behaving like a whore.

The reciprocity and equality of the conjugal debt is also challenged by the many examples of medieval women seeking release from the obligations of sexual intercourse. Confessional writers indicated that women were sometimes 'frigid' or refused to fulfil the debt when asked because they feared the dangers of pregnancy and childbirth that were the consequences of intercourse without contraception. Other writers acknowledged that in practice a husband should be sensitive to his wife's needs and refrain from sexual demands if she were ill or had a headache. Religious considerations also might result in women seeking permanent release from the obligations of the conjugal debt. For example, in the early fifteenth century, Margery Kempe was but one of many married women who tried to convince their husbands to forego marital sexual relations. After giving birth to thirteen children, Margery finally convinced her husband to abstain from sex and release her from the obligations of the conjugal debt, although it was clear that, at least early in her marriage, Margery had enjoyed their sex life. She remarks on her belief that she and her husband 'had often . . . displeased God by their inordinate love, and the great delight that each of them had in using the other's body'.[14]

Medieval belief in the linkage between conception and sexual pleasure also led to unexpected social consequences. For example, the twelfth-century theologian, William of Conches, wrote that 'prostitutes who have sexual relations for money alone and who take no pleasure during the act have no emission and thus do not conceive'. Prostitutes were supposed not to become pregnant with their clients, but would with a husband or lover. Similarly, and more insidiously, this belief was applied to victims of rape. If the woman became pregnant as a result of assault, it was believed she must have experienced sexual pleasure. As William of Conches observed: 'Although in rape the act is distressing to begin with, at the end, given the weakness of the flesh, it is not without its pleasures'.[15]

Experiencing the sexual body

Although medical and pastoral writers alike affirmed the theory that both men and women must achieve orgasm and their seeds mingle for conception to occur, there is little evidence to suggest what happened in practice. While medical texts discuss explicitly the techniques to ensure sexual gratification, neither moral treatises nor imaginative literature indicate that kissing and foreplay or the pursuit of sexual pleasure, especially for women, were particularly prevalent; indeed, kissing and other forms of sexual play were repeatedly condemned as obscene. In the late seventh century, the penitential of Theodore forbade couples from seeing each other naked, a prohibition repeatedly stressed in sermons and confessional treatises throughout the Middle Ages. How this injunction might have translated into medieval marital life remains problematic. Manuscript illustrations from the late twelfth century onwards regularly portray couples in bed nude, occasionally wearing only nightcaps. Similarly, illustrations of brothels and bathhouses portray naked men and women together in bathtubs. Even patients in hospitals are portrayed two to a bed and naked, although no spousal, familial or erotic relationship was implied in such a situation.

With or without foreplay, 'immoderate' pleasure, even during conjugal, procreative sex, was consistently condemned in prescriptive literature. This moral anxiety was based on the church father Jerome's assertion, in his highly misogynistic but influential treatise, *Against Jovinian*, that: 'Nothing is more obscene than to love your own wife as if she were your paramour'. Given the necessity of orgasm for procreation, what 'immoderate' pleasure may have meant remains perplexing, but a moralist like Thomas of Chobham, writing in the early thirteenth century, provides one possible interpretation. He condemned husbands who had trouble controlling their sexual urges and who constantly sought to have sexual relations with their wives. Such men, he opined, were consumed by lust and harassed their wives day and night, as if they were in bed. This, he concluded, was sinning through lascivious kisses and filthy embraces. Yet, reprehensible as this behaviour was, it was also why the conjugal debt was necessary; lest those uncontrollable urges be directed outside the conjugal relationship and result in adultery or, even worse, sodomy.

There are very few sources we can juxtapose to the didactic and prescriptive texts written by the theoretically celibate clergy. How sexually active medieval people might have thought about themselves as sexual beings or how they experienced sex for the most part remains beyond our gaze, owing to the paucity of self-reflective writing by members of the laity. For this reason, the evidence of the famous love affair between the philosopher Peter Abelard and the learned Heloise is of particular value. Although their words were written for public circulation rather than to express private feelings, and however tempered the sentiments and memories were by the intervening events, their letters nevertheless provide a glimpse of passionate, erotic desire and a satisfying sexual relationship.

The letters written between Abelard and Heloise detail, in a he said/she said form, the outlines of their relationship. The very fact that they had a relationship is revealing of sexual practices in twelfth-century towns. Abelard, the young academic and man about town, successfully seduced his brilliant pupil, Heloise. Not only did this couple begin a sexual relationship outside of marriage, but Abelard also suggests that Heloise was not the only young woman who would have considered such a relationship: '[A]t that time I had youth and exceptional good looks as well as my great reputation to recommend me, and feared no rebuff from any woman I might choose to honour with my love'.[16] This boast was corroborated by Abelard's enemy, Fulk of Deuil, who, after Abelard had been castrated, suggested that men no longer needed to fear that he would seduce their wives and daughters.

Abelard, moreover, reveals that he and Heloise were something of sexual adventurers: 'In short, our desires left no stage of love-making untried, and if love could devise something new, we welcomed it. We entered on each joy the more eagerly for our previous inexperience, and were the less easily sated'.[17] Even when formally rejecting and condemning their earlier relationship as wanton and evil, the joy and pleasure of their sexual encounters is still evident: 'So intense were the fires of lust which bound me to you that I set those wretched, obscene pleasures, which we blush even to name, above God as above myself'.[18] Of equal significance, however, is Heloise's view of their sexual relationship. She consistently indicates that she was an active participant in their sexual encounters: '[W]e enjoyed the pleasures of an uneasy love and abandoned ourselves to fornication'.[19] Her memories, too, are filled with sexual desire:

> In my case, the pleasures of lovers which we shared have been too sweet –
> they can never displease me, and can scarcely be banished from my thoughts.
> Wherever I turn they are always there before my eyes, bringing with them
> awakened longings and fantasies which will not even let me sleep.[20]

One of the fascinating discrepancies between the accounts of Abelard and Heloise is how they present male and female roles in sexual activity. Abelard presents himself as the virile sexual actor, initiating sex, even forcing it on a passive and hesitant Heloise. 'Even when you were unwilling, resisted to the utmost of your power and tried to dissuade me, as yours was the weaker nature I often forced you to consent with threats and blows'.[21] This is quite different from Heloise's view – she indicates she was both a willing participant and a recipient of sexual pleasure. It is she who still longs for Abelard's touch, a sentiment notably absent from his letters. How we account for such a discrepancy reveals much about the challenges of studying sex in past societies. Was Abelard slipping into literary convention? Was Heloise simply adopting the language of her male-authored literary models? Or do these two differing accounts indicate two differing perspectives? Could Abelard's words be repeating the stereotypes of female sexual passivity and male sexuality as active, even aggressive, whereas Heloise was expressing her lived experience as a woman who had known sexual pleasure? There is currently no sure way to resolve such a conundrum.

Setting the boundaries of natural sex

Just as science and theology reinforced each other in terms of understanding the world, God's creation and how it worked, so, too, sex and the social and religious controls over its exercise were understood with reference to nature. Sexual acts were categorized according to whether they were natural or unnatural. Natural sex acts were few, indeed, whereas the category of unnatural sex acts was elastic and almost infinitely expandable. According to medieval theologians and moralists, the only natural sex acts occurred between one man and one woman, in the so-called 'missionary' position, with the man emitting semen in the woman's vagina. For such a natural act also to be without sin, the man and woman were required to be united in legitimate marriage and to be engaging in sex for the purpose of procreation, rather than to experience sexual pleasure. All other sexual activities were condemned as contrary to nature (*contra naturam*). This meant that for most of the Middle Ages, Church officials condemned alternative heterosexual positions for intercourse, perhaps because they might be intended to enhance the couple's pleasure. For this reason, too, sexual experimentation, such as that acknowledged by Abelard and Heloise, was officially discouraged. In the thirteenth century, theologians and moralists increasingly recognized occasional exceptions to these proscriptions. For example, Raymond of Pennafort permitted alternative positions in order to protect the fetus if the woman were pregnant (although sex during pregnancy was considered sinful) and Alexander of Hales allowed them if necessitated by illness or obesity. Albert the Great discussed four alternate positions for intercourse: side-by-side, seated and standing were all blameworthy but only intercourse from behind was reckoned to be a mortal sin. It went without saying that oral sex acts were utterly prohibited, contrary to nature and mortally sinful, as was coitus interruptus, both because it was contraceptive and because it involved the spilling of semen outside the 'natural vessel'.[22]

When examined on a wider scale, the moralized concept of nature dominated, defined and directed the exercise of sex and sexuality into a narrow range of behaviour and object choice. By the mid-thirteenth century, Thomas Aquinas, building on Augustine's foundational categories, refined the definition of acts contrary to nature under the rubrics of: unnatural heterosexual positions, masturbation, sodomy[23] and bestiality. While all these activities were mortally sinful, sodomy and bestiality were particularly heinous and subject to prosecution in both ecclesiastical and secular courts. Thus, sexual excess or sexual deviance was no longer only sinful within the private forum of the confessional, but also was a crime that could be prosecuted and punished in the public forum. This hardening of attitudes towards sexual sins is the culmination of an increasingly harsh attitude towards sexual activity throughout the preceding centuries.

In the penitential literature from the fifth to the eighth centuries, sexual sins were discussed in rather formulaic terms and, although there is little consistency between them, their authors tended to evaluate sexual sins as relatively minor transgressions to judge by the mildness of the penalties that were recommended. In early Irish penitentials, even bestiality was regarded as scarcely more serious than masturbation, and oral and anal sex were punished no more severely (and often less severely) than fornication or adultery. In the early eighth century, the penitential of Theodore advised seven years of penance for a man who engaged in oral sex with a man ('who puts semen in the mouth'), noting that other penitentials would require lifelong penance for fellatio.[24]

Penitentials also tended to prescribe milder penances for younger people who engaged in sinful sex acts. For example, young boys who experimented with sodomy received half the penance of an adult man. Men who engaged in such acts occasionally might be assigned penance for half as long as those who were habitual sodomites. Similar distinctions were made for women who engaged in same-sex sexual activity. If a woman were unmarried the penance was much lighter than if she were married or a nun. Other penitentials distinguished the relative sinfulness of women engaging in sexual activities together involving rubbing or manual stimulation from those that employed 'diabolical instruments', that is, dildos. The use of devices that would mimic penetration by a penis was far more serious because it so clearly mimicked male activity and usurped men's natural function.[25] For the most part, though, the penitentials did not focus on sex acts between women or female masturbation. There was a sense in which, in the absence of a penis and without the ejaculation of semen, such activities were ultimately negligible transgressions of both the natural and moral order.

Gradually, sexual sins came to be considered not only more serious by the Church, but also acts that could be punished in the public forum or by secular authorities. In the thirteenth century, secular law codes began to include punishments such as dismemberment or burning for men found guilty of sodomy. This roughly coincided with Aquinas's characterization of unnatural acts, all of which were subsumed under the category of sodomy. This expansion of the category of sodomy at once served to render previously minor invisible sins, such as masturbation or heterosexual experimentation, as serious as habitual anal intercourse. It also made same-sex acts between women, hitherto neglected or minimized, as serious as male/male sodomy and a capital offence that could be prosecuted in secular courts.

Sex: unnatural and dangerous

In the course of the twelfth and thirteenth centuries, it is possible to identify increasing social anxiety surrounding sex in general, and the distinction between natural and unnatural acts in particular. For example, the evaluation of bestiality underwent a profound transformation between the early and later Middle Ages.[26] In patristic writing and in the penitentials there are indications that humans and animals were viewed as having profoundly different natures. Animals were property to be owned and objects not worthy of a great deal of consideration. So, in this early period, bestiality was reckoned to be roughly equivalent to masturbation. Columbanus prescribed six months' penance for each of these sins, compared to three years for fornication and ten years for sodomy. In the eighth- and ninth-century penitentials, bestiality is considerably more serious, garnering penances equivalent to those of sodomy. Gradually, the animal came to be considered more of a partner in sin, rather than a handy object to be used for sinful pleasure. As a result, animals, as participants, were equally subject to punishment for their part in the sinful act. Many later penitential and confessional manuals advised that the animal should be killed and its flesh and hide burned.

By the twelfth century, however, there was a more profound transformation in the views of human and animal interaction. Writers such as Gerald of Wales began to offer accounts of monstrous births of creatures that were the half-animal/half-human offspring of the unnatural commingling of man or woman and beast. This suggests a blurring of the boundaries that were perceived to separate humans from animals. No longer profoundly different and non-threatening, animals were now both a threat and a temptation to humans. Even more frightening than the blurring of the lines of demarcation between human and animal, rational and irrational, was the gradual linkage of animals with demons. In the thirteenth century, there was an increasing fear of and belief in the possibility of demons having sex with humans – demons, moreover, who often appeared partially or wholly in the shape of animals. Thus, bestiality was no longer merely a sexually transgressive act that involved an animal, as it might any other malleable object. Rather, bestiality was now linked with a frightening supernatural world in which evil could assume the forms of common beasts and intercourse with animals could just as easily be intercourse with demons, the blurring of species mirroring the blurring of the natural and unnatural realms.

All of the activities gathered under the rubric of unnatural acts overthrew the natural order and the moral order. Inherent in bestiality was the rejection of rational sex for irrational, and the elevation of inferior animals to sexual partners. This is why Aquinas considered it to be the most heinous of unnatural acts. Masturbation was a rejection of the natural procreative intent of sexual intercourse, by the ejaculation of semen outside 'the natural vessel'. Alternative heterosexual positions implied a rejection of the natural order of the sexes which mandated women be inferior to men. Thus, the woman-on-top position was seen as a metaphor for female social domination. Sex from behind was even worse, since it mirrored the position of animals mating and thus implied a rejection of human superiority over animals. Sodomy, too, was a rejection of the natural and moral order, not only because it involved same-sex partners, but also because in the process one partner assumed an inappropriate gender role.

The notion that men were active and women passive in the sex act was ubiquitous in medieval belief. A variety of discourses and contexts mutually reinforced this evaluation. From medicine to theology, from popular culture to high culture, this was the prevalent view of the sexual order. Sex between women, then, challenged this paradigm because it implied that one woman usurped the active role. Male writers assumed that a sex act between women would involve some kind of phallus substitute. If it did not, it was an inferior sex act and one less threatening to the established order because the male principle was absent. As a result, there are frequent references to a differential in punishment for the active versus the passive female partner. In the example of sex between women, however, it is less certain whether it was the sex act per se which was being punished or the assumption of male prerogatives. Similarly, the evaluation of sodomy assumed that one man lowered himself to the passive, female, penetrated role. Thus, it was not just the non-procreative sex act that was punished, but also the rejection of the natural order that required that men be superior sexual actors.

Given the gender-based stereotypes, the rigid moral code and the strict social organization which formed the context for same-sex relations, it is difficult to move beyond proscription and towards an understanding of the psychological and emotional dimensions of same-sex relations in medieval society. Indeed, it is largely owing to this world-view and to the formulaic nature of the Latin texts that many scholars of sexuality have concluded that in the Middle Ages there were only sex acts as opposed to internalized sexual orientations or sexual identities as we understand them now. Nevertheless, despite the challenges posed by the sources and the medieval context, it is occasionally possible to glimpse the human being and the emotions hidden behind the acts. The example of Arnaud de Verniolle is particularly useful in this regard.

Arnaud de Verniolle was a cleric who was arrested and interrogated by the Inquisition, as part of the major proceedings against the Albigensians of Montaillou in the 1320s. Arnaud was prosecuted for a combination of sodomy and heresy. The records of his interrogation hold a wealth of information about the language of sexual relations between men and the self-conception held by Arnaud and the men with whom he engaged in sex.[27] The transcripts reveal Arnaud (and the clerk who recorded and translated his testimony) struggling to express concepts and feelings which were beyond the reach of the available language. Just as Heloise struggled to find a way to express her own experience of sexuality in a language that gendered sexual activity masculine, so Arnaud endeavoured to find a way to express his own sexuality in a language that was constrictingly heterosexual. Moreover, in Arnaud's testimony, there is evidence of 'sexual orientation as life practice'[28] rather than as mere bodily acts.

Arnaud asserted that, although he might be termed a sodomite, he had never engaged in anal sex, either as active or passive partner. Rather, the preferred sexual act between Arnaud and his paramours was side-by-side interfemoral stimulation. Although Arnaud was in his thirties and his lovers in their late teens, these men did not conform to active and passive roles dictated by age. Rather, they assumed different positions in turn, revealing a far more egalitarian approach to sex acts between men than the classical model of an active older man dominating/penetrating a passive younger man. Moreover, Arnaud explicitly rejected the prone face-to-face position of heteronormative intercourse, indicating that rather than mimicking sex between a man and a woman, he and his partners engaged in sex side-by-side, which, he believed, did not threaten the natural and moral order. In Arnaud's testimony, his linguistic frustration is evident, as he endeavoured to find a way to understand himself and to express his identity and his love of men in the absence of a vocabulary that would allow for it.

The overthrowing of the natural order expressed by unnatural sexual acts linked sex with that other sin that overthrew the order of creation, heresy. This was apparent in the perceived linkage between bestiality and sex with demons in animal or quasi-animal guise. The ties between unnatural sex acts and demons were perceived to be many, particularly in the thirteenth and fourteenth centuries. Even masturbation, that solitary vice that could be deemed to harm no one else, was rendered more dangerous and horrifying through a link to the supernatural world. Caesarius of Heisterbach, for example, fanned the flames of fear by asserting that, as a result of masturbation and other illicit seminal emissions, 'demons collect all wasted seed, and from it fashion for themselves human bodies, both of men and women, in which they become tangible and visible to men'.[29] Thus, far from being an innocuous, private sin, masturbation was transformed into a means to facilitate the work of demons and their nefarious goal to seduce and corrupt humans.

A more problematic and extreme example of the increasing anxiety that was associated with all sexual activity, whether normative and legitimate or unnatural and sinful, is evident in the popular stories about the phenomenon of *penis captivus*.[30] This was the belief that during intercourse either the woman's vaginal muscles could constrict in a spasm or the man's penis could swell out of proportion, rendering the couple unable to extricate themselves from the sex act. They were thus helplessly joined until discovered by others who released them through prayers of intercession. A series of these stories, stretching from the early Middle Ages through the fourteenth century, focuses on couples who were supernaturally locked together as punishment for engaging in sexual activity which polluted sacred spaces. In early versions, the couple were in a cemetery or churchyard, in later versions, in the church itself and even in the sanctuary. In eleventh- and twelfth-century versions, the couple were adulterous, fornicating or even committing sacrilege, with the man identified as a monk or priest. In later stories, however, the adhering couple were married. They were, then, engaged in marital relations which, according to the doctrine of the conjugal debt, could be required whenever and wherever demanded. The notion of sex in a church was not as improbable as it might first appear. There were few places in medieval towns and villages that provided couples, whether married or not, with privacy. Records reveal couples had sex in fields, ditches, the forest, the cemetery and, yes, in the church.

The tales of the locked couples are more than absurd stories or urban legends (enduring into the late twentieth century). They were also a means by which society could neutralize some of the danger inherent in sexual activity, which was exacerbated by the imperatives of the conjugal debt. Semen, like blood, had a polluting effect, not only for people but for spaces. Just as murder and the shedding of blood defiled a church, so, too, did the emission of semen. But, according to canon law, the conjugal debt had to be rendered if demanded, even in a consecrated space. The problem was that if this happened secretly, the community would be unaware its church had been desecrated and needed to be ritually cleansed and reconsecrated. The miracle stories of the adhering couples reveal deeply rooted fears of secret, sexual pollution. The stories gave people some assurance that, should secret pollution occur, God would find a way to reveal it so that ritual cleansing could occur.

For medieval people, sex was as powerful as it was complex. Ideas about the sexed body oscillated wildly from being the temple of the lord to a sign of the fall and the locus of original sin. Sexual intercourse allowed for the propagation of believers, while also revealing humanity's irrational, lower appetites. Mystics described the soul's union with God in the erotic language of orgasm in a period when people believed women could be seduced in their beds by demons. Sex, in its many and various meanings, was complicated and imbued with danger. From the vantage point of the postmodern world, sex in the Middle Ages ought to be viewed as complicated, mysterious and perhaps even inexplicable in some of its aspects. While many medieval beliefs about sex may appear to be transparent, even familiar, other aspects remain virtually inexplicable. What, for example, was the meaning and function of the phenomenon of the penis tree? Various media, including a French manuscript illustration, a carving on a wooden casket and a fresco on a fountain in Italy, portray trees whose robust fruit of erect penises is being harvested by women. On a basic level, the motif invokes sex and fertility, but surely it is more complicated than that. Surely, within the matrix of medieval beliefs, values and practices, the portrayal of the penis tree had a complex and multivalent meaning which remains thus far hidden from curious postmodern eyes. If this example can readily be perceived to be opaque and not immediately explicable, it should also serve as a reminder to those who study the medieval history of sex to beware of the temptations of the apparent familiarity and immutability of human sex and sexuality. The web of meanings that entwine sex, even in its most recognizable forms, veils and transforms, rendering the obvious obscure.

Guide to further reading

Ariès, Philippe and Béjin, André (eds), *Western Sexuality. Practice and Precept in Past and Present Times*, trans. A. Foster (Oxford, 1985).

Atkinson, Clarissa. '"Precious Balsam in a Fragile Glass": The Ideology of Virginity in the Later Middle Ages', *Journal of Family History* 8 (1983), pp. 131–43.

Baldwin, John W., *The Language of Sex. Five Voices from Northern France around 1200* (Chicago, IL, 1994).

Boswell, John, *Christianity, Social Tolerance, and Homosexuality. Gay People in Western Europe from the Beginning of the Christian Era to the Fourteenth Century* (Chicago, IL, 1980).

Boswell, John, *Same-Sex Unions in Premodern Europe* (New York, 1994).

Bottomley, Frank, *Attitudes to the Body in Western Christendom* (London, 1979).

Brooten, Bernadette J., *Love Between Women. Early Christian Responses to Female Homoeroticism* (Chicago, IL, 1996).

Brown, Peter, *The Body and Society. Men, Women, and Sexual Renunciation in Early Christianity* (New York, 1988).

Brundage, James A., *Law, Sex, and Christian Society in Medieval Europe* (Chicago, IL, 1987).

Brundage, James A., *Sex, Law and Marriage in the Middle Ages* (Aldershot, 1993).

Bullough, Vern L., 'On Being a Male in the Middle Ages', in Clare A. Lees (ed.), *Medieval Masculinities. Regarding Men in the Middle Ages* (Minneapolis, MN, 1994), pp. 31–45.

Bullough, Vern L. and Brundage, James A., *Sexual Practices and the Medieval Church* (Buffalo, NY, 1982).

Bullough, Vern L. and Brundage, James A., *Handbook of Medieval Sexuality* (New York, 1996).

Burger, Glenn and Kruger, Steven F. (eds), *Queering the Middle Ages* (Minneapolis, MN, 2001).

Cadden, Joan, *Meanings of Sex Difference in the Middle Ages. Medicine, Science, and Culture* (Cambridge, 1993).

Elliott, Dyan, *Spiritual Marriage. Sexual Abstinence in Medieval Wedlock* (Princeton, NJ, 1993).

Elliott, Dyan, *Fallen Bodies. Pollution, Sexuality, and Demonology in the Middle Ages* (Philadelphia, PA, 1999).

Flandrin, Jean-Louis, *Le sexe et l'Occident, Évolution des attitudes et des comportements* (Paris, 1981).

Foucault, Michel, *The History of Sexuality*, trans. R. Hurley (3 vols, New York, 1978–86).

Fradenburg, Louise and Freccero, Carla (eds), *Premodern Sexualities* (New York, 1996).

Jacquart, Danielle and Thomasset, Claude, *Sexuality and Medicine in the Middle Ages*, trans. M. Adamson (Princeton, NJ, 1988).

Jordan, Mark D., *The Invention of Sodomy in Christian Theology* (Chicago, IL, 1997).

Karras, Ruth Mazo, *Common Women. Prostitution and Sexuality in Medieval England* (New York, 1996).

Karras, Ruth Mazo, 'Sexuality in the Middle Ages', in P. Linehan and J.L. Nelson (eds), *The Medieval World* (London, 2001), pp. 279–93.

Lemay, Helen Rodnite (ed.), *Homo Carnalis. The Carnal Aspect of Medieval Human Life* (Binghamton, NY, 1990).

Murray, Jacqueline, 'On the Origins and Role of "Wise Women" in Causes for Annulment on the Grounds of Male Impotence', *Journal of Medieval History* 16 (1990), pp. 253–49.

Murray, Jacqueline and Eisenbichler, Konrad (eds), *Desire and Discipline. Sex and Sexuality in the Premodern West* (Toronto, 1996).

Partner, Nancy F., 'No Sex, No Gender', *Speculum* 68 (1993), pp. 419–43.

Payer, Pierre J., 'Early Medieval Regulations Concerning Marital Sexual Relations', *Journal of Medieval History* 6 (1980), pp. 353–76.

Payer, Pierre J., *Sex and the Penitentials. The Development of a Sexual Code 550–1150* (Toronto, 1984).

Payer, Pierre J., *The Bridling of Desire. Views of Sex in the Later Middle Ages* (Toronto, 1993).

Salisbury, Joyce E. (ed.), *Sex in the Middle Ages. A Book of Essays* (New York, 1991).

Sautman, Francesca Canadé and Sheingorn, Pamela (eds), *Same Sex Love and Desire Among Women in the Middle Ages* (New York, 2001).

Notes

1   J. Boswell, *Christianity, Social Tolerance, and Homosexuality. Gay People in Western Europe from the Beginning of the Christian Era to the Fourteenth Century* (Chicago, IL, 1980).
2   M. Foucault, *The History of Sexuality*, vol. 1, *An Introduction*, trans. R. Hurley (New York, 1978).
3   '[S]hould she/Be more than woman, she would get above/All thought of sex, and think to move/My heart to study her, and not to love'. John Donne, 'The Primrose', in A.L. Clements (ed.), *John Donne's Poetry* (London, 1992), p. 39.
4   T. Laqueur, *Making Sex. The Body and Gender from the Greeks to Freud* (Cambridge, MA, 1990).
5   J. Murray, '"Law of sin that is in my members": The problem of male embodiment', in S.J.E. Riches and S. Salih (eds), *Gender and Holiness. Men, Women and Saints in Late Medieval Europe* (London, 2002), pp. 14–15.
6   *The Letters of Abelard and Heloise*, ed. B. Radice (Harmondsworth, 1974), p. 133.
7   *Bawdy Tales from the Courts of Medieval France*, trans. and ed. P. Brians (New York, 1972), pp. 24–36.
8   Gerald of Wales, *Jewel of the Church*, ed. J.J. Hagen (Leiden, 1979), 2.11, p. 166.
9   This tale, recounted by William of Newburgh, is discussed in E.J. Kealey, *Medieval Medicus. A Social History of Anglo-Norman Medicine* (Baltimore, MD, 1981), p. 42.
10 Gerald of Wales, *Jewel of the Church*, 2.11, pp. 166–7.
11 Andreas Capellanus, *The Art of Courtly Love*, trans. J.J. Parry (New York, 1941), p. 199.
12 *The* Historia Pontificalis *of John of Salisbury*, ed. and trans. M. Chibnall (Oxford, 1986), pp. 7 and 14–15.

13 Dyan Elliott, 'Bernardino of Siena versus the Marriage Debt', in J. Murray and K. Eisenbichler (eds), *Desire and Discipline. Sex and Sexuality in the Pre-modern West* (Toronto, 1996), pp. 168–200.

14 *The Book of Margery Kempe*, trans. B.A. Windeatt (London, 1985), p. 46.

15 Cited in D. Jacquart and C. Thomasset, *Sexuality and Medicine in the Middle Ages*, trans. M. Adamson (Princeton, NJ, 1988), p. 64.

16 *Letters of Abelard and Heloise*, p. 66.

17 *Letters of Abelard and Heloise*, pp. 67–8.

18 *Letters of Abelard and Heloise*, p. 147.

19 *Letters of Abelard and Heloise*, p. 130.

20 *Letters of Abelard and Heloise*, p. 133.

21 *Letters of Abelard and Heloise*, p. 147.

22 J.A. Brundage, 'Let Me Count the Ways. Canonists and Theologians Contemplate Coital Positions', *Journal of Medieval History* 10 (1984), pp. 81–93.

23 In this context, sodomy refers to same-sex sexual activity involving two men or two women rather than to the more specific activity of anal intercourse.

24 P.J. Payer, *Sex and the Penitentials. The Development of a Sexual Code 550–1150* (Toronto, 1984).

25 J. Murray, 'Twice Marginal and Twice Invisible: Lesbians in the Middle Ages', in V.L. Bullough and J.A. Brundage (eds), *Handbook of Medieval Sexuality* (New York, 1996), pp. 191–222.

26 J.E. Salisbury, 'Bestiality in the Middle Ages', in J.E. Salisbury (ed.), *Sex in the Middle Ages. A Book of Essays* (New York, 1991), pp. 173–86.

27 A translation of the testimony of Arnaud and his partners can be found in M. Goodich, *The Unmentionable Vice: Homosexuality in the Later Medieval Period* (Santa Barbara, CA, 1979), pp. 93–123.

28 F.C. Sautman, '"Just Like a Woman": Queer History, Womanizing the Body, and the Boys in Arnaud's Band', in G. Burger and S.F. Kruger (eds), *Queering the Middle Ages* (Minneapolis, MN, 2001), pp. 168–89.

29 Caesarius of Heisterbach, *The Dialogue on Miracles*, trans. H. von E. Scott and C.C. Swinton Bland (2 vols, London, 1929) 2, p. 12; 1, p. 139.

30 D. Elliott, 'Sex in Holy Places. An Exploration of a Medieval Anxiety', *Journal of Women's History* 6 (1994), pp. 6–34.

8

Gender and femininity in medieval England

Cordelia Beattie

In December 1395 some London officials saw a couple 'lying by a certain stall in Soper's Lane', having sexual intercourse, and went to arrest them. It was discovered that the person in women's clothing was one John Rykener, calling himself Eleanor. The other man, John Britby, claimed that he had seen 'Eleanor', 'dressed up as a woman, thinking he was a woman, asking him as he would a woman if he could commit a libidinous act with her'. 'Eleanor' had asked for payment and they had gone to the stall where they were caught. This story is confirmed by Rykener and the subsequent interrogation focused on who had taught him to exercise this vice, for how long, in what places and with what persons. The interrogation is summarized in Latin in the city's Plea and Memoranda Rolls.[1] It emerges that not only had men had sex with Rykener 'as with a woman', but Rykener had also had sex with women 'as a man'.[2] This atypical case offers a useful way into the subject of medieval gender as it allows us to put to one side our categories of men/women and to think instead about what qualities and activities were thought to pertain to men/women when they were being 'as a man' or 'as a woman'. That is, it allows us to think about gender, rather than sex.

In the first part of this essay, I shall use the example of John/Eleanor Rykener to discuss different theories of gender. In particular, I shall examine the relationship between sex and gender, and show how different theorists' views of gender might be applied to this medieval example. The second part of the essay will focus on gender theory and medieval femininities. The key aspect will be gender as a strategy for reading, and how this might reveal alternative versions of femininity which might not be seen if we just looked for 'women'. It will be argued that femininity differs from masculinity in a number of ways, not just in qualities and characteristics, but as a system. Throughout this essay, though, it will be clear that an understanding of masculinity should inform our understanding of femininity.

What is 'gender'?

The use of 'gender' as a category of historical analysis is an evolving part of the study of history and, as such, is an exciting one in which to participate. The word 'gender' is used by historians and others in a variety of ways, not always carefully thought through – it might just be a euphemism for 'sex', as it is increasingly used on the forms that we fill in, or shorthand for 'women'. The latter use perhaps stems from seeing women as the marked 'gender', with more in common as a group than men. As a result, some readers assume that 'gender' is just a feminist code word for women and their historical complaints about social injustice. However, gender offers an important new cultural vocabulary, one that lets us see the pervasive ideas about females *and males* that are evident in all social and cultural constructs, most especially historical evidence in all its forms. For example, it might be used to discuss cultural and social norms that are associated with the 'masculine' and the 'feminine', as distinct from biological differences between the male and the female.

These are some of the examples that are most easy to spot, but 'gender' has been understood and used in a wide variety of ways. For example, Joan Scott – in a seminal essay – has argued that 'gender is a primary way of signifying relationships of power', even when 'concepts of power . . . are not literally about gender'.[3] One example she gives is of middle-class reformers in nineteenth-century France who depicted workers as subordinate, weak and sexually exploited – terms coded feminine; terms coded masculine included producers, the strong and protectors. Such formulations naturalize disparities in power between men and women. The significance of gendered language is an issue to which I shall return. The intention of this essay is not to provide a fixed meaning of gender; writers will continue to use the word in a variety of ways. A discussion of some of the key uses, though, may help readers to spot the theoretical assumptions that underlie different scholars' approaches in the future.

Sex/gender

Among the common uses of the term 'gender', that of gender as cultural – in opposition to sex as biological – is held by many historians as an underlying assumption. Although this contention has now been challenged (as we shall see), it is still worth exploring its history. The English language distinction between sex and gender was first developed in the 1960s by psychiatrists and psychologists studying intersexed and transsexual patients. The argument was that such patients suffered from a deep sense of disconnection between their biological or bodily sex and their psychological gender. An American psychoanalyst, Robert Stoller, publicized these distinctions in his 1968 book, *Sex and Gender: On the Development of Masculinity and Femininity*. In his understanding, gender was considered an aspect of individual identity, as a psychosocial construct which referred to how people categorized themselves as well as to how they were seen by others. Stoller developed the concept of 'gender identity' to refer to a person's psychological experience of belonging to one sex or another.[4] In using a figure such as John/Eleanor Rykener to discuss gender, I am therefore following in a long-standing tradition.[5] By studying cases of those who experienced a mismatch between their sex and their gender, Stoller devised a theory of what was normative. Atypical examples can be good to think with because they highlight and make visible what otherwise just merges with the social scenery. However, for historians, the question is less about what made this one man dress as a woman (only this single record has come to light about Rykener), but more about what his actions and the reactions of others reveal about social norms. Indeed, when feminists picked up on this distinction between sex and gender in the 1970s, the latter was understood as dominant gender norms.

Feminists adopted the distinction between sex and gender as a way of rejecting biological determinism or what is known as essentialism (that is, that what it is to be a woman is biologically determined). If what it meant to be a woman varied according to society and culture, then the position of women could be both studied and changed in the present. A key essay was Gayle Rubin's 'The Traffic in Women' (1975). Rubin used the expression 'the sex/gender system' as a synonym for 'patriarchy', understood as social forces that oppress women. She defined it as 'the set of arrangements by which a society transforms biological sexuality into products of human activity'. Here, sex is biological, the raw material for gender, while gender represents the oppressive social norms brought to bear on sexual difference.[6] Rubin's understanding of gender was influenced by her aim: she wanted to explain why and how women's oppression was maintained in widely different cultures. This blurring of the topics 'gender' and 'women' is something that continues to this day, but it is increasingly realized that dominant gender norms affect men, too (see Derek Neal, below, chapter 9).

Rykener, gender identity and gender norms

The switch from individual 'gender identity' (a person's qualities and ways of being as male or female) to dominant gender norms (or sexual stereotypes) is not just a historical one; these are still the two key ways in which gender is used in gender theory and they can be demonstrated with reference to John/Eleanor Rykener. As the record is written from the perspective of London's civic authorities, it is perhaps easier to see their presumptions about gender, that is, dominant gender norms. While wearing a dress obviously did not make Rykener a woman, the case does allow us to think about what qualities and activities were thought to pertain to women in late fourteenth-century London.

First, it was clearly not acceptable for a man to wear 'women's clothing', although from other incidents that Rykener recalls it is evident that while wearing women's clothes he could pass as a woman.[7] Second, as regards sexual activity, it seems that the 'male' was seen as the 'active' partner, the 'female' as passive:

> Rykener further confessed that [he] went to Beaconsfield and there, as a man, had sex with a certain Joan . . . and also there two foreign Franciscans had sex with him as a woman . . . Rykener further said that he often had sex as a man with many nuns and also had sex as a man with many women both married and otherwise . . . Rykener further confessed that many priests had committed that vice with him as with a woman.

The legal record gives Rykener's responses to questions he was asked rather than an unprompted admission; medieval confessional literature shows a similar concern about sex with those in religious orders and the married. However, leaving to one side the types of person Rykener claims he had sex with, it is noticeable that, according to the record, the men had sex with him, whereas he had sex with the women.[8] It can be argued that both these examples relate to what Rubin called oppressive gender norms: men were constrained in their choice of clothing, as were women;[9] women should not be 'active' sexual partners – men had to be the sexual initiators. However, both examples – dressing as a woman, having sex as a man or as a woman – also relate to acts; it was what Rykener *did* which made people see him as a man or a woman. This theory of gender as performative is one that arose from dissatisfaction with the sex/gender distinction.

If biological 'sex' is seen as the basis on which gender difference is constructed then gender always follows sex. This was undoubtedly the normative premise that the psychiatrists and psychologists in the 1960s were working with when they argued that intersexed and transsexual patients suffered from a discrepancy between their sex and their gender. In the 1990s, the idea that gender automatically followed from sex was seen by some poststructuralist theorists as worryingly essentialist. This led Judith Butler to argue that there could be more than two genders and produce her theory of gender as performative.[10] This has been an influential theory but understood in a variety of ways. In the preface to the tenth-anniversary edition of her 1990 book, *Gender Trouble*, Butler acknowledges that she uses the term 'performative' in two key ways. First, that we create gender by our expectation that there is a gendered interior essence. Second, that this is not a one-off act of creation: we are all constantly performing our gender in a way that produces our gender identity. Gender identity, for Butler, is actually created by a sustained series of acts which we carry out (although this is not the same as taking on a gender *role* as a choice).[11] In Rykener's case, then, we could argue that his actions create his gender identity. It is at this point, if we follow Butler rather than Stoller, that gender norms and gender identity perhaps collide in our reading. The record presents Rykener as switching between acting as a man and acting as a woman because certain acts were seen as gender specific by the society in which he lived. However, did Rykener compartmentalize his life in this way or did he see both facets as integral to his own gender identity? Answering the latter question would take us into the realms of speculation, given that the only record we have is written from the perspective of London's civic authorities.[12]

While the record of the interrogation does dwell largely on sexual acts, it also suggests that Rykener was able to pass as either a man or a woman in everyday life, which would have involved other gendered behaviour. The authorities were less interested in Rykener passing as a man, which is accepted as the norm, but presumably his relations with the women mentioned involved more than the act of sexual intercourse; there is no suggestion that he had sex with the women as a prostitute. We are given more detail, though, about Rykener's passing as a woman:

for five weeks before the feast of St. Michael's last [he] was staying at Oxford, and there, in women's clothing and calling himself Eleanor, worked as an embroideress; and there in the marsh three unsuspecting scholars . . . practiced the abominable vice with him often. . . . John Rykener further confessed that on Friday before the feast of St. Michael [he] came to Burford in Oxfordshire and there dwelt with a certain John Clerk at the Swan in the capacity of tapster for the next six weeks, during which time two Franciscans . . . committed the above-said vice with him.

It seems that Rykener lived as a woman while in Oxfordshire, working first as an embroideress ('*in arte de* brouderer') and then as a barmaid ('*in officio de* tapster'). While men could embroider and sell ale, these are both occupations usually practised by women in late medieval England. Rykener is, therefore, more than a man who wears female clothing and has sex with men. He also claims to have lived and worked as a woman. This perhaps suggests a deep feminine identification on his part.

As regards the embroidery, we can perhaps speculate further as to how Rykener entered this trade. The woman who is said to have first dressed him in women's clothing is one Elizabeth Brouderer. Elizabeth's byname, 'Brouderer', suggests that she was an embroiderer and it seems likely that she was the same Elizabeth who, in 1385, was accused in London's courts of taking in girls as apprentices in the craft of embroidery and getting them to work as prostitutes.[13] Rykener alleges that Elizabeth also prostituted her daughter, but at night, without light and then in the morning showed the men Rykener 'dressed up in women's clothing, calling him Eleanor and saying they had misbehaved with her'. He also says that a rector 'had sex with him as with a woman in Elizabeth Brouderer's house'.[14] However, as Rykener then went to Oxford and worked as an embroiderer, it seems that he did also learn this craft. The record is more explicit about who taught him to practise the 'detestable vice in the manner of a woman': one Anna, 'the whore of a former servant of Sir Thomas Blount'. As with the embroidery, though, what else did Rykener learn from these women that the record omits? Surely passing as a woman involves more than wearing women's clothes and doing conventional women's work.

We need to think about how people learn their gender roles. As we have seen, for Butler there is no interior essence of gender. It is not necessary, therefore, to be a transvestite to 'perform' gender; the women in this case are also 'performing' their gender. However, even for Stoller – whose idea of core gender identity is a very early pre-verbal recognition that one has been assigned a correct sex/gender match ('yes, I, a female, am a girl') – gender is seen as proceeding from sexual identity in a *gradual learned process*, beginning at birth ('It's a boy!') and expanding as the infant encounters the social world. The psychological core of identity just makes it easier for males to aspire to meet the expectations of masculine character and females to accept or adapt to the feminine idea. Whichever understanding of gender one adopts – Butler's, Stoller's or the more general view that it is the cultural and social interpretation of biological difference – gender is still something that is acquired. So, how did the women that Rykener met learn what it was to be a woman in fourteenth-century England? What the exact expectations were for males and females differ from one society to another and are rarely consistent within a society. The next section will therefore focus on some of the variety of ideas about being feminine in medieval England.

What is 'femininity'?

Masculinity/femininity

In this volume, the topic of gender has intentionally been split into two chapters: one on femininity and one on masculinity. While the subject of 'medieval women' has a longer scholarly pedigree than that of 'medieval masculinity', masculinity as a term is more commonly used by historians than femininity. This is perhaps partly political; some of the objections to 'gender history' which have come from within 'women's history' are rooted in convictions about the wider implications of emphasizing 'women' as a subject of history.[15] But it also perhaps speaks to the differences between masculinity and femininity. If masculinity and femininity are the social and cultural understandings of the biological categories male and female, there is no need to expect them to be constructed in similar fashions. Gender is made up of what people thought (or assumed or presumed or accepted as true or what ought to be true) about males and females, including themselves. It is not about accurate observation of what males and females were actually like.

In all gender systems that we know much about, femininity was never contested, sought or paraded in the way that masculinity was (however defined).

Masculinity is something to be achieved and men can be unsuccessful – be rendered effeminate – but femininity is not something to strive for or something that can be lost easily as it could be both good and bad and equally 'feminine'. Women were not measured by how much of it they 'had'. Masculinity always was (and is) a contested, desired, feared and anxiety-generating something. On this issue of the instability of masculinity, Robert Shoemaker and Mary Vincent comment:

> According to feminist psychoanalytic theory the transformation of boys into men is a more difficult process than that of girls into women due to the fact that boys must reject the early 'feminine' influence of their mother, and that this forces men to work harder to assert and defend their masculinity. This argument may seem unprovable and unhistorical, but in fact historically in Europe far greater effort has been devoted to the socialization of young men than young women in homosocial environments such as journeymen's associations, schools and youth groups: boys are frequently expected to pass some kind of test (such as a school initiation rite) in order to be considered 'real' men (with those who failed branded 'effeminate').[16]

Even if we reject the psychoanalytic reading for why men need to assert their masculinity in ways that do not hold true for women and femininity, the difference can perhaps be explained with reference to social and political structures.[17] Pre-modern European societies were patriarchal in the sense that women had a lesser degree of wealth, status and power than men of their own class. All women were, therefore, subordinate to some men through all life-cycle stages and careers, whereas there was an expectation that some men would prove themselves dominant over others, and as a result, gradations of dominant and subordinate masculinities were constructed.[18]

Masculinity and femininity, then, are not analogous systems, but they do coexist and impinge on each other. In medieval texts, as well as in contemporary culture, there are references to male individuals being 'less of a man' or 'super men'. As the Shoemaker and Vincent quotation points out, the successful are manly to an extreme, whereas those who do not score highly in their masculinity are thought 'effeminate'. There is notably less movement on the femininity scale, however. For women, success can lead to the status of 'honorary man' or 'virile', as some of the medieval texts put it, but they do not ever seem threatened with a sub-feminine status. That unsuccessful men might slip into femininity, whereas successful women can encroach on masculinity, also suggests a hierarchy of systems.[19] Of course, this is what one might expect in a culture that privileges the male.

But what do we mean by 'femininity'? For some, this probably has connotations of behaving and dressing 'in a feminine way'. But for others it might mean what women are basically like, in mainly unflattering ways connected to emotionality and bad character traits like envy and vanity. It also had different specific meanings in the medieval period. So, how do we uncover medieval views of femininity? As historians, our only access to the past is through the surviving evidence – for medievalists, this is textual. We can use a range of sources, for example, court records (as I did with Rykener), scientific treatises, sermons, conduct books and saints' Lives. However, as well as thinking about who has written each text and for what purposes, we also need to be aware that each text might be the product of competing views, that is, competing *discourses*. The Foucauldian understanding of discourse has been summed up by Joan Scott as 'a historically, socially, and institutionally specific set of statements, terms, categories, and beliefs'. Chris Weedon, discussing Foucault, also notes that 'Discourses represent political interests and in consequence are constantly vying for status and power'.[20] When thinking about gender and medieval society, then, we need to think about how we can *read* gendered ideas in the surviving texts and how much social authority to impute to these ideas.

Reading for gender

I shall demonstrate some of the different ways we can read for gender with three different texts. The first is the fourteenth-century English translation of Bartholomew the Englishman's Latin encyclopaedia, *On the Properties of Things*. Although probably written on the continent in about the 1240s, it was popular into the sixteenth century. It makes use of many earlier authorities, such as the scientific treatises of Aristotle, other classical texts, the Bible and its glossators such as St Augustine, as well as later commentators, and so can be used to suggest many of the misogynistic views of women that circulated in medieval texts. I shall examine parts of Book 6, on the ages of man, to address the perceived relationship between masculinity and femininity, as well as different models of femininity. This will allow us to move away from a monolithic idea of femininity, both in the sense of how we think about it and in terms of recognizing that medievals did not have a single model either. It also enables us to go beyond the idea of gender as a fixed set of traits or roles that necessarily follow from a particular sex, to a recognition of variations in masculinity and femininity.

In Book 6, after a section on infants, Bartholomew has separate sections on boys and girls. In the chapter on girls, he cites Aristotle as his authority for how physiological differences between the sexes suggest differences in character:

> Aristotle says that every woman generally has wavier and softer hair than a man, and a longer neck. A woman's complexion is fairer than a man's and her face is cheerful, gentle, bright and amiable. . . . She has a swift wit. She is merciful but also envious, bitter, deceitful, easily led, and quick to the

pleasures of Venus. . . . A woman is more gentle than a man (she weeps more easily than a man), and is more envious and more amorous. And there is more malice in a woman than a man. She has a weak nature, tells more lies, is more modest, and slower in working and moving than a man.[21]

Here, man is the norm against which the woman is measured: she is, for example, 'more gentle', 'weeps more easily' and has 'more malice'. Aristotle is cited here as categorizing women in both positive and negative ways. The negatives do predominate, though, and some are given double emphasis, for example, as when women are deceitful and tell more lies, are more amorous and 'quick to the pleasures of Venus', and are bitter and more malicious. However, the scientific, matter-of-fact tone does not make the chapter sound merely like a diatribe against women.

In a subsequent chapter on man, the register shifts to a more polemical one and we encounter a picture of the good woman and the bad woman rather than the monolithic woman. Bartholomew cites his source here as Fulgencius, a fifth- to sixth-century grammarian, sometimes identified as the bishop of Ruspa, who in his sermon on the marriage of Cana likened Christ to the good man, the holy church to the good wife and the synagogue to the evil wife. Here the passage is used to discuss the various attributes of a good and bad wife:

no man is more wretched nor has woe and sorrow than he that has an evil wife, crying, jangling, chiding and scolding, drunken and unsteadfast and contrary to him, costly, showily dressed, envious, destructive, and leaping over lands and countries and thieving, suspicious, and wrothful. . . .

In a good spouse and wife these conditions are needed: that she be busy and devout in God's service; meek and obedient to her husband, and fair speaking and gracious to her household; merciful and good to wretches that be needy; kind and peaceable to her neighbours; ready, aware, and wise in things that shall be avoided; rightful and patient in suffering; busy and diligent in her doing and deeds; modest in clothing; sober in moving; wary in speaking; chaste in looking; honest in bearing; dignified in going; shamefast among the people; merry and glad with her husband; and chaste in private.[22]

The good wife is the opposite of the evil wife: meek and wary in speaking, rather than nagging and gossiping; chaste, modest in dress and sober in behaviour, rather than extravagant in her dress and 'leaping over lands and countries', getting up to no good; also obedient to her husband rather than quarrelsome. In the picture of the good wife, then, we perhaps have a model of ideal femininity. However, if we look at the qualities that each possesses, both the good wife and the evil wife can be seen as typically feminine. If we compare it with the Aristotelian passage on girls, again there is the idea of women as merciful and modest, but also as envious, bitter and deceitful. This is not to say that we should construct a binary model of femininity, rather than a binary between femininity and masculinity. Rather, we should recognize that there was no single medieval idea of femininity. If women are silent they are feminine, if they are talkative they are feminine. This brings us back to my earlier point that it is easier for women to attain femininity and harder to escape it than seems to be the case with men and masculinity.

A detailed examination of Bartholomew's encyclopaedia could show how different discourses are selected to support one another. However, for my purposes it is enough that this first example suggests how one text might make use of different discourses and that there might be different models of femininity. My second approach, building on the idea of plural discourses, is to look for medieval ideas of gender that the texts did not mean to reveal. I have mentioned dominant and subordinate masculinities, but not dominant and subordinate femininities. It has been argued that subordinate masculinities are 'subjected to greater repression than their feminine counterparts', because women, already dominated by men, were not as difficult to control socially.[23] However, in a society in which the Church held up virginity as the highest ideal, but marriage and motherhood – in addition to ensuring the continuation of family lines – were integral to households which managed working practices and were agents of government, it is likely that there would have been different articulations of ideal femininity. What we can look for in a text, then, are the moments in which it is revealed that the dominant discourse is not the only one, that it has achieved its dominance by asserting itself against an implicit competing discourse.[24] We can see an example of a clash of ideals in the fifteenth-century *Book of Margery Kempe*, which primarily concerns the devotional life of a medieval Englishwoman. It is at such a point, where the text engages with this conflict, that it perhaps reveals a bit more than it intended.

The eponymous Margery is said to have lived in late fourteenth- and early fifteenth-century Norwich. She was the daughter of a five-times mayor of Lynn and can be associated with a mercantile elite. The book narrates the life of this medieval woman, who was pulled between her role as a wife, with responsibilities towards her husband, and her desire to be a full-time religious witness, which entailed a life of celibacy.[25] At a crucial juncture in this life, and so in the book, there is the following scene between Margery and her husband, John:

It happened one Friday, Midsummer Eve, in very hot weather – as this creature was coming from York carrying a bottle of beer in her hand, and her husband a cake tucked inside his clothes against his chest – that her husband asked his wife this question: 'Margery, if there came a man with a sword who would strike off my head unless I made love with you as I used to do before, tell me on your conscience – for you say you will not lie – whether you would allow my head to be cut off, or else allow me to make love with you again, as I did at one time?'

. . . And then she said with great sorrow, 'Truly, I would rather see you being killed, than that we should turn back to our uncleanness.'

And he replied, 'You are no good wife.'[26]

In this scene, Margery is said to be – by her husband – 'no good wife'. The 'good wife' in Bartholomew's encyclopaedia was equally obedient to please her husband and devout in God's service. In Margery's case these two standards have come into conflict and the latter has won out.

To put the incident in context, the first chapter of *The Book of Margery Kempe* opens with marriage, childbirth and post-partum illness (described as 'tormented with spirits').[27] The book then sets out how her unsuccessful business ventures made her see the error of her ways and she 'forsook her pride, covetousness, and the desire for worldly dignity, and did great bodily penance, and began to enter the way of everlasting life as shall be told hereafter'.[28] Part of this process was a desire for chastity, but her husband refused; under Christian rules, married partners both owed each other the so-called 'conjugal debt' of sexual intercourse, and therefore neither party could become chaste without the other's consent. As part of a pact with Christ, Margery is told that if she fasts from both meat and drink on a Friday, her wish will be granted. This comes true as, further on in the same chapter, Margery's husband says he will grant her this desire if she will grant him three things: share a bed with him, pay his debts before going on pilgrimage to Jerusalem, and eat and drink with him on Fridays. Christ tells her that she can give in on the latter request as he only asked her to fast so that she could get her husband's consent. Margery therefore agrees to the last two of her husband's wishes but asks that he never comes to her bed again and he agrees.

In this example, the importance of chastity triumphs over wifely obedience. In an ecclesiastical model of hierarchical chastity, virginity was prized most, followed by vowed chastity in widowhood, followed by chastity in marriage (which meant reserving sex for the purposes of procreation). However, in giving the husband's point of view so that Margery can state her own case, the text perhaps allows us a glimpse of a third role for Margery, that of economic provider. That he wants her to share his bed and eat with him fits with the role of submissive wife, but asking her to pay his debts does not. As a married woman, Margery's property should be his already.[29] The text does not dwell or elaborate on how the husband wants Margery to pay his debts; indeed, the debate focuses on the other two requests, one granted after a conversation with Christ, one refused. However, that

Margery controls enough money to pay debts that her husband himself could not pay hardly fits the conventional models of wifely behaviour. Here, though, it seems to be accepted unproblematically by Margery, her husband and the scribe of the text, since no one comments on it as unusual. Historians who have looked at this text make much of an earlier chapter in which Margery describes her work, first as a brewer and then running a mill.[30] The purpose of these descriptions within the text seems to be to illustrate Margery's conversion experience and warn about the dangers of pride. Their inclusion fits a conventional model, whereas paying her husband's debts does not.

This is not to say that women did not do such jobs or earn money. Women in the medieval West were expected to perform tremendous amounts of responsible and difficult work, at every social level, and were expected to handle their household, financial, business and estate-management responsibilities with good judgement and prudence. While we can see elements of this in the description of the good wife in Bartholomew's encyclopaedia – she should keep herself busy with the work of the household – these realities tended to be under-expressed in cultural terms, that is, they did not alter the centre of woman-defining ideas which tended to a static adherence to denigrating gender models. In general, in the medieval societies that we know about, chastity (symbolic purity or the opposite, sexuality) had a strong and pervasive presence in *all* cultural constructions of femininity. In this chapter of *The Book of Margery Kempe*, the debate between Margery and her husband revolves primarily around chastity versus marital sexuality, but the importance of his debts (and, by implication, her money) suggests that, in a society in which women did fulfil vital roles in the household economy, other models of femininity or female-appropriate behaviour were available in society, even if our access to them is limited. All through crucial parts of this text, neither Margery nor the scribe seem to have found anything worth much attention in Margery as a businesswoman and as an independent controller of money whose uses she decides. All this points to aspects of female-appropriate behaviour that the conventional models of femininity seem to have no room for and so just ignore. We need to look out, then, for those moments in which they inadvertently surface.

My final example continues the theme of conflict between the ideals of chastity and marriage, and returns to the questions of gender norms and the relationship between masculinity and femininity. I shall use a couple of passages from the twelfth-century *Life of Christina of Markyate* to demonstrate what gender as a strategy for reading can reveal, which we might not see if we just looked for 'women'. The coded gendered language used suggests the different judgements attached to qualities and actions and how they might vary according to discursive purpose.

We are told that Christina made a vow of virginity as a young girl. As in the Lives of the virgin martyrs, it is not enough to *be* a virgin to achieve heroic sanctity, but that virginity must be under threat and still preserved. In such stories, the virgin is often out of step with contemporary mores (the Lives of the virgin martyrs are set in pagan times), so we would expect to find opposition to her proposed way of life from family and important officials. It is in their responses that we are given insights into how society might have viewed a woman rejecting traditional gender roles such as that of wife.

In the *Life*, a bishop, whose advances Christina had spurned, conspired to have her married to a young man named Burthred. Although technically married, Christina refused to consummate the marriage. Her parents let the husband into her room at night, 'in order that . . . he might suddenly take her by surprise and overcome her'. The wary Christina met him, dressed and awake, and 'strongly encouraged him to live a chaste life, putting forward the saints as examples'.[31] She talks with him for much of the night in this way. When Burthred leaves the room, the responses towards him suggest how a woman's rejection of her proper wifely role could affect the masculinity of the men associated with her:

> When those who had got him into the room heard what had happened, they joined together in calling him a spineless and useless fellow. And with many reproaches they goaded him on again, and thrust him into her bedroom another night, having warned him not to be deceived by her long speeches and glowing words nor to be made effeminate [*effeminetur*]. Either by force or entreaty he was to gain his end. And if neither of these sufficed, he was to know that they were at hand to help him: all he had to mind was to act the man [*modo meminerit esse virum*].[32]

In this passage, Christina is still constructed as feminine; as we have seen, in Bartholomew's encyclopaedia, women are thought to be deceitful. However, Burthred is in danger of becoming effeminate by his failure to get Christina to submit to his will. When he first entered her room he was told to 'overcome her' ('oppresse'), and now he is advised to add entreaty to force as long as he is successful.[33] He is told 'to act the man', to be a man. As in the Rykener example, this text implicitly suggests the idea of performing gender. If Burthred does not have sex with Christina he will lose his manliness, and Christina's refusal to fulfil her feminine role as dutiful wife is emasculating him.[34] While the text lets us see this, it is also firmly on the side of Christina's chosen vocation and privileges her alternate feminine model, that of the heroic holy maiden.

Further on in the *Life*, about eight years later, we have another scene of sexual temptation, but here Christina is gendered masculine. Christina, now an anchoress, was sent to live with a cleric, but the devil tries to overcome their chastity:

> Sometimes the wretched man, out of his senses with passion, came before her without any clothes on and behaved in so scandalous a manner that

I cannot make it known . . . And though she herself was struggling with this wretched passion, she wisely pretended that she was untouched by it. Whence he sometimes said that she was more like a man than a woman, though she, with her more masculine qualities [*virago virtute virili*], might more correctly have called him effeminate [*effeminatum*].[35]

Again we have the man rendered 'effeminate' by Christina's actions, whereas Christina – usually described as a virgin (*virgo*) – is elevated to the position of *virago* (literally a heroic or warlike woman). The cleric is seen as feminized for his failure to control his passion, whereas, in the previous example, Burthred was rendered effeminate by his failure to assert his sexual will. Christina is seen as 'manly' here for having the moral strength to resist the seducer she feels passionately about. In the previous example, she was gendered feminine, by her parents' warning to Burthred about her deceptive words, and by the writer for preserving her virginity under threat. These differences, within and between the examples, allude to my earlier points that masculinity and femininity are not analogous systems, that masculinity is seen as superior and that what characteristics pertain to which gender might vary according to discourse and discursive purpose. In these passages, conventional ideas of who is feminine and who is masculine slide around with the author using whatever ideal or insult serves his purpose at that point.

The study of medieval gender, then, is not just about knowing the stereotypical gender notions that prevailed in medieval society. Medieval people were not so simple-minded that their cultural clichés say everything there is to know about them, any more than our clichés reveal everything about us. In order to use 'gender' as a category of historical analysis, we need to pay attention to competing discourses, what texts inadvertently reveal and the presence of coded gender language. A reading of historical sources which pays attention to gender in this way can add depth and insight to historical analysis. While texts like that of the cross-dressing prostitute with which I started provoke a reaction by upsetting our conventional views of medieval society, it is not only such examples that can benefit from a gendered reading.

Guide to further reading

For a more detailed account of the history of the sex/gender debate, including a helpful discussion of Judith Butler's arguments, see Toril Moi, *What Is a Woman? And Other Essays* (Oxford and New York, 1999), ch. 1, esp. pp. 3–59. For a brief summary of current doubts about a sex/gender distinction, also from the disciplines of biology, anthropology and psychology, see

Merry E. Wiesner-Hanks, *Gender in History* (Malden, MA, and Oxford, 2001), pp. 2–5.

For the importance of gender as a historical tool, see Joan Wallach Scott's 'Gender: A Useful Category of Historical Analysis', *American Historical Review* 91 (1986). However, note the preface to the revised edition of her *Gender and the Politics of History* (2nd edn, New York and Chichester, 1999) and the new essay which it contains, 'Some More Reflections on Gender and Politics', as her views have changed.

For the arguments of Robert J. Stoller, a good place to start is his *Presentations of Gender* (New Haven and London, 1985), ch. 2. For a recent account by a medievalist which refutes the applicability of psychoanalytic 'truths' (and also challenges the contention that masculinity is always more at risk than femininity), see Ruth Mazo Karras, *From Boys to Men: Formations of Masculinity in Late Medieval Europe* (Philadelphia, PA, 2003), pp. 3–12.

For the theories of Judith Butler, a good starting point is her 'Imitation and Gender Insubordination', in *Inside/Out: Lesbian Theories, Gay Theories*, ed. Diana Fuss (New York, 1991). For a discussion of her ideas by a medievalist, see Allen J. Frantzen, 'When Women Aren't Enough', *Speculum* 68 (1993), pp. 455–7; this issue was devoted to the topic of medieval gender and the other essays in it are all worth reading.

Shannon McSheffrey's discussion of gender, which includes references to Scott and Butler, is very clear: *Gender and Heresy: Women and Men in Lollard Communities 1420–1530* (Philadelphia, PA, 1995), pp. 2–5. Another medievalist with a sophisticated understanding of gender is Sharon Farmer. See her Introduction to *Gender and Difference in the Middle Ages*, eds Sharon Farmer and Carol Pasternack (Minneapolis, MN, and London, 2003), pp. ix–xv.

Notes

1   Both the edited Latin text and the English translation are the work of David Lorenzo Boyd and Ruth Mazo Karras, 'The Interrogation of a Male Transvestite Prostitute in Fourteenth-Century London', *GLQ: A Journal of Lesbian and Gay Studies* 1 (1995), pp. 461–5. The full record can also be seen – in translation, in Latin and in a facsimile of the document itself – on the Internet Medieval Sourcebook: Paul Halsall, 'Medieval Sourcebook: The Questioning of John Rykener, A Male Cross-Dressing Prostitute, 1395', May 1998, http://www.fordham.edu/halsall/source/1395rykener.html (accessed 14 August 2004).

2   Translations in this paragraph (and elsewhere) are from Boyd and
     Karras, 'Interrogation of a Male Transvestite Prostitute', pp. 462–3.
3   Joan W. Scott, 'Gender: A Useful Category of Historical Analysis',
     American Historical Review 91 (1986), p. 1069.
4   Robert J. Stoller, *Sex and Gender: On the Development of Masculinity
     and Femininity* (London, 1968); chapter 7 includes part of his earlier
     article, 'A Contribution to the Study of Gender Identity', *International
     Journal of Psychoanalysis* 45 (1964), pp. 220–6.
5   For other discussions of Rykener, see Carolyn Dinshaw, *Getting
     Medieval: Sexualities and Communities, Pre- and Postmodern* (Durham,
     NC and London, 1999), ch. 2, and Judith Bennett, 'England: Women
     and Gender', in *A Companion to Britain in the Later Middle Ages*, ed.
     S.H. Rigby (Oxford and Malden, 2003), pp. 87–8 and 99–101 (the latter
     was only available to me after the drafting of this essay).
6   Gayle Rubin, 'The Traffic in Women: Notes on the "Political Economy"
     of Sex', in *Toward an Anthropology of Women*, ed. Rayna R. Reiter
     (New York and London, 1975), p. 159. Rubin, though, did recognize sex
     as 'a social product' (p. 166) and, by talking of the sex/gender system
     rather than just gender, she did not adhere to the strict separation that
     others adopted.
7   I use 'pass' here to get across the notion that Rykener does more than
     cross-dress. Shibanoff uses it to refer to 'complete, even if episodic,
     assumption of the appearance of the opposite sex', in contrast with
     'cross-dressing', which is 'partial masquerade' only: Susan Shibanoff,
     'True Lies: Transvestism and Idolatry in the Trial of Joan of Arc', in *Fresh
     Verdicts on Joan of Arc*, eds Bonnie Wheeler and Charles T. Wood (New
     York and London, 1996), p. 57, n. 33. I would contend that 'the
     appearance of the opposite sex' includes the assumption of gender
     roles. As we do not know Rykener's gender identity, he might also be
     said to 'pass' as a man.
8   The binaries of male/active and female/passive have also been found in
     confessors' manuals. See J. Murray, 'The Absent Penitent: The Cure of
     Women's Souls and Confessors' Manuals in Thirteenth-century
     England', in *Women, the Book, and the Godly: Selected Proceedings of
     the St Hilda's Conference, 1993*, vol. I, eds L. Smith and J.H.M. Taylor
     (Cambridge, 1995), p. 20.
9   For reactions to a woman wearing men's clothing, see Shibanoff, 'True
     Lies'.
10   Another move has been to challenge the status of 'sex' as 'natural', as
     opposed to the cultural gender. It is argued that science does not stand
     outside culture and that our 'biological facts' are also products of a par-
     ticular society and subject to reinterpretation. See, for example, Gisela
     Bock, 'Women's History and Gender History: Aspects of an International
     Debate', *Gender and History* 1 (1989), pp. 10–15.

11 Judith Butler, *Gender Trouble: Feminism and the Subversion of Identity* (New York and London, rev. edn, 1999), pp. xiv–xv and 33.

12 As Karras and Boyd state, 'there is no way of verifying the veracity of Rykener's account': Ruth Mazo Karras and David Lorenzo Boyd, '"*Ut cum muliere*": A Male Transvestite Prostitute in Fourteenth-Century London', in *Pre-modern Sexualities*, eds Louise O. Fradenburg and Carla Freccero (New York), p. 102.

13 *Women's Lives in Medieval Europe: A Sourcebook* (New York and London, 1993), pp. 211–12. Karras and Boyd also make this identification: '"*Ut cum muliere*"', p. 114, n. 22.

14 This incident is also intriguing in that Rykener claims that, having stolen two of the rector's gowns, he avoided returning them by threatening the man that he 'would make [his] husband bring suit against him'. This appears to be another example of Rykener passing as a woman in that he adopts the married woman's covered legal position (on this, see Caroline M. Barron, 'The "Golden Age" of Women in Medieval London', *Reading Medieval Studies* 15 (1989), pp. 35–7). It has been pointed out to me that Rykener could have tried a different tactic and revealed himself to be 'John'; that Rykener appears to choose, voluntarily, a 'feminine' strategy perhaps suggests feminine identification on his part: conversation with Dr John H. Arnold, regarding thoughts generated by students in his 'Gender in the Middle Ages' course, Birkbeck College, University of London (2003).

15 See, for example, Mary Evans, 'The Problem of Gender for Women's Studies', in *Out of the Margins: Women's Studies in the Nineties*, eds Jane Aaron and Sylvia Walby (London, 1991).

16 Robert Shoemaker and Mary Vincent, 'Introduction. Gender History: The Evolution of a Concept', in *Gender and History in Western Europe*, eds Robert Shoemaker and Mary Vincent (London, 1998), pp. 5–6.

17 For a brief account of how Stoller's concept of core gender identity revises Freud's argument that femininity is more problematic, see Robert J. Stoller, *Presentations of Gender* (New Haven, CT, and London, 1985), pp. 14–18.

18 On dominant and subordinate masculinities, see R.W. Connell, *Masculinities* (Cambridge, 1995), pp. 76–81; for the medieval period, see especially D.M. Hadley, 'Introduction: Medieval Masculinities', in *Masculinity in Medieval Europe*, ed. D.M. Hadley (Harlow, 1999), pp. 4–13.

19 Clover found what she calls a one-sex, one-gender system in Norse society – womanishness was contemptible for men and women, extraordinary women could become 'social men' – but argues that it was very different from later systems: Carol Clover, 'Regardless of Sex: Men, Women, and Power in Early Northern Europe', *Speculum* 68 (1993).

20 Joan W. Scott, 'Deconstructing Equality-versus-Difference: Or, The Uses of Poststructuralist Theory for Feminism', *Feminist Studies* 14 (1988), p. 35; Chris Weedon, *Feminist Practice and Poststructuralist Theory* (2nd edn, Oxford, 1997), p. 40.

21 *Love, Marriage, and Family in the Middle Ages: A Reader*, ed. Jacqueline Murray (Peterborough, Ontario, 2001), pp. 448–9.

22 *On the Properties of Things: John Trevisa's Translation of* Bartholomæus Anglicus De Proprietatibus Rerum. *A Critical Text*, ed. M.C. Seymour *et al.* (3 vols, Oxford, 1975–88), I, p. 309 (I have modernized the English).

23 Shoemaker and Vincent, introduction, p. 6, citing David D. Gilmore, *Manhood in the Making: Cultural Concepts of Masculinity* (New Haven, CT and London, 1990), p. 221.

24 Steven Justice has written about this in respect of chronicle accounts of the Peasants' Revolt of 1381 in his *Writing and Rebellion: England in 1381* (Berkeley, CA and Los Angeles, CA, 1994), p. 7.

25 Much has now been written on how far the Margery of the text is a construction of the scribe who wrote it, following an earlier draft and Margery's dictation, e.g. Lynn Staley, 'Margery Kempe: Social Critic', *Journal of Medieval and Renaissance Studies* 22 (1992), pp. 159–84.

26 *The Book of Margery Kempe*, trans. B.A. Windeatt (Harmondsworth, 1985), p. 58.

27 *The Book of Margery Kempe*, trans. Windeatt, p. 41.

28 *The Book of Margery Kempe*, trans. Windeatt, p. 45.

29 Barron, 'Golden Age', pp. 35–7.

30 *The Book of Margery Kempe*, trans. Windeatt, ch. 2.

31 *The Life of Christina of Markyate*, ed. and trans. C.H. Talbot (Oxford, 1959), p. 51.

32 *The Life of Christina of Markyate*, ed. and trans. Talbot, pp. 50–3 (I have slightly amended the translation).

33 *The Life of Christina of Markyate*, ed. and trans. Talbot, p. 50.

34 Foyster has argued regarding men in early modern England that 'a man's sexual activities, or lack of them, were central to notions of honourable and dishonourable manhood': Elizabeth A. Foyster, *Manhood in Early Modern England: Honour, Sex and Marriage* (London and New York, 1999), p. 10.

35 *The Life of Christina of Markyate*, ed. and trans. Talbot, pp. 114–15 (I have slightly amended the translation).

9

Masculine identity in late medieval English society and culture

Derek Neal

Introduction: bodies and discourses

Gender is, in many ways, about language, and this is significant since the word was originally a grammatical term, serving to classify nouns and pronouns. Speakers of English have a disadvantage in seeing this because our language, unlike many others, has lost grammatical gender. It does not classify inanimate objects, countries or abstract concepts as masculine, feminine or neuter. Nevertheless, it is not so long since we still commonly referred to ships and nations, in English, as 'she', and young children still often think of dogs as 'boys' and cats as 'girls', in ways that defy practical reasoning. Indeed, children are expert in the practice of 'gendering' the world without reference to actual sex; their awareness of a distinction between 'boys' and 'girls' pre-dates, in developmental terms, their understanding of anatomical difference between male and female. They grasp gender before they understand sex.[1]

So gender is inseparable from our primary means of comprehending the world and our place in it. This was even truer of the Middle Ages. In English, 'man' could mean a male adult, as it does now; it could also mean, generally, 'human being', a connotation that only became uncommon in the late twentieth century. More strikingly, 'manhood' could mean 'the qualities and characteristics of adult males', either actual or ideal; it could mean 'the state of being adult and male'; or it could mean 'humanity', the quality of being human, or (substantively) 'the human race', as we might put it now. Though grammatically genderless, the English language still conflates maleness with a default setting for humanity – the baseline norm. If the category 'man' overlapped with 'human being', with all the virtues that distinguished humans from the rest of Creation, little was left that was positive for the category 'woman' to contain. Ideas and values keyed to sexual difference (manly/womanly, male/female, masculine/feminine) thus inevitably cast the feminine in terms of lack, insufficiency and subordination. This value system, enshrined in language, has been sufficiently effective to endure in discourse today.

To study masculinity is not the same thing as to study men, but one cannot divorce the two, especially from the historian's perspective. Somewhere behind all the material from the past left to us, there once were human beings. To do history means to confront several levels of reality, difficult sometimes to reconcile: the human source and the textual residue. Historians do their work through discourse: the documentary sources that inform us consist, usually, of text, so that when we apprehend and analyse them, everything we do becomes subject to the many processes of transformation distinguishing language. No meanings are fixed and since we can't physically contact the people who produced the sources, they have no reality to us outside the text. Even the uncommon historians who rely largely on non-textual sources, such as art historians or those who use cumulations of numerical data, still must use linguistic discourse at some point in formulating their ideas, because there is no other way to tell their stories. (Similar problems crop up even in interviewing a living person.)

However, to regard our historical subjects simply as abstracted textual imprints not only denies them their humanity – an error in itself – but also makes us run the risk of missing something important in our analysis: the degree to which human experience is grounded in the body, in the physical and in meanings derived from it. Our perception of the world depends on our bodies; both pleasure and pain reside in them; they provide the metaphors by which we understand things outside ourselves at all levels of complexity. Thinking that this has ever not been the case is foolish, and in any society as dependent on the rhythms and vicissitudes of the natural world as pre-modern Europe, the reach of the physical into the realm of the mental and psychological must surely have been deeper.

Our understanding of gender as language, as a discourse, ought to help, not obscure, our understanding of people in history as real: as human beings who lived full and complex lives. Theorists of gender, influenced by poststructuralism, have long emphasized gender's tenuous connections to sexed bodies. In this they show the influence of the feminist initiative to unhinge anatomy (or 'biology') from destiny, to sever the connections between the form of a body and the social roles it then performs. They have taken this to some confusing extremes, and it is often difficult to see what is meant by insisting that 'the body' is a 'product' of discourse or 'created' by it. What follows is an analysis of a historical text, to make this clearer.

Torke's pintle

In 1470, the young Norfolk gentleman, John Paston II, who was serving in the army of King Edward IV, wrote a letter to his brother, John III, in which he mentioned offhandedly:

> Item, there is come a new little Torke, which is a well-visaged fellow of the age of 40 year, and he is lower than Manuell by a handful and lower than my little Tom by the shoulders, and more little above his pap [nipple]. And he hath, as he said to the King himself, 3 or 4 sons [for] children, each one of them as high and as likely as the King himself. And he is legged right enough, and it is reported that his pintle [penis] is as long as his leg.[2]

Obviously this fellow caught Paston's attention and he thought him worth mentioning in a letter: not as an important matter, just an interesting tidbit. And why was it so interesting? First, Torke was unusually short in stature, shorter than other small men Paston knew. Paston tries to give his reader a picture of the man's body, whose oddness may be a matter of proportion; the description divides his upper torso into sections, and it seems he was particularly compressed between his shoulders and chest. The language used makes it difficult to imagine the body precisely. Still, Torke was not hideously deformed, but a 'well-visaged fellow', pleasant enough to look at; moreover, to serve in an army, even as a servant, he could not have been too much disabled.

Rhetorically, the description of Torke's abnormalities serves to set up the real surprise, the contrast with what was (surprisingly) normal about him. Despite his shortness, Paston reports, he had fathered several children, male children at that, who were of normal, in fact admirable stature, as tall and 'likely' (impressive, strapping) as the king. (Other observers reputed Edward IV to be a tall and, at this point, good-looking man, still in his late twenties in 1470.) That Torke's smallness was a matter of the upper body seems supported by Paston's comment that he was 'legged right enough' – that his legs seemed normal, at least in proportion. Like a good storyteller, Paston saves the sauciest bit for last: the rumour that Torke's genital endowment made up for whatever was lacking in the rest of his body. That last clause is admittedly an interpretation on my part, but I think it's supported by the text. Paston didn't begin his story by saying, 'there's a guy in our company who is hung like a stud horse', which would have changed the relative importance of all the other elements. As it is, Torke's penis appears, a rumour and punchline, as the final bit of proof: of what is both most important and most unusual. It does this by establishing emphatically that this is a man's body: both male and masculine. A gender-analysis précis of the passage might say: 'Torke is a tiny man, with a weirdly proportioned body, and therefore we might expect him to fall short [pun intended], to be weak and unmanly, to be unable to carry himself as a man in our society among other men. Yet actually, he has several sons, normal and attractive sons, so he's obviously sexually capable and socially capable, in that he has secure heirs. And he didn't hesitate to say all this to the prime alpha male of them all, the King himself. So consequently, his penis must be big enough to fill the gaps, to make the equation balance'. Note the roles of both body and discourse in this example. The body is (sort of) easy to see. What are the discourses that bear on masculinity in history, on our meaning-making?

The most obvious relation is that between the printed (originally written) text – Paston's letter – and the writing of our/my interpretation of it. Yet both of those depend on and intersect with other discourses. The discourse of letter-writing, ordinary family communication about very practical and prosaic matters, is what allowed this passage to become known, riding alongside more 'important' information in the terms of writer (John II) and reader (John III). So one writer–reader relation (Paston to us) encloses another (John II to John III). And the fifteenth-century one is a discourse between men; it is a homosocial discourse. The reported discourse, the rumours about Torke and his self-presentation, are homosocial as well. They belong to a system of communication between men in an all-male context (the military cohort), one of many such contexts where talk established reputation. Not only was Torke's body, or most of it, on display for public comment, but so were his interactions with other men. What he said to the king, and, by implication, other men who outranked him, then formed part of another discursive network, which affected the way his peers and superiors saw him. Competition, and making and holding one's place in a masculine hierarchy, depended on discourse (language, speech), not just on physical conflict or sheer social and monetary advantage: a point driven home forcefully by other kinds of evidence.

Discourse bears on gender in this passage in another way, which may be difficult to grasp at first. It follows from the précis I suggested above. In the imaginative logic Paston enters and sustains, Torke's penis – the thing that marks him most securely as physically male – is an outcome, a conclusion, of the profile he has cut through checking all the right masculine boxes: husband, father, public figure (in his own small way), competitor – therefore, sexually complete man as well. Here we have a small example of what 'discursive creation' might mean in a real life. Masculinity is a set of meanings, broader ones enfolding smaller ones. We can state the broadest as 'the meaning of being male in a given society'. Still, that meaning has to reside somewhere; it only exists in reference to something, or someone. The frame of reference of gender, and particularly of masculinity, is very broad. Whose meaning is it? We cannot see masculinity, or femininity, as something inherent: as something that individuals possess, as a trait or characteristic inside themselves. This goes a little further, or at least in a different direction, from the common admonition not to see gender as 'essential' or 'natural', as an automatic outcome of biological sex. However, neither is gender wholly extrinsic: a set of rules (boys are to be like this, girls are to be like that) which individuals must obey or bear the consequences. It is instead, in the words of one non-historian, a 'compromise formation': a negotiation between individuals and their cultures, a dynamic that ultimately becomes part of their identities – their selves.[3] Seeing gender as a set of meanings underscores that it is something made, produced, even in a sense achieved (achievement is a particularly important idea for masculinity); that making and becoming happen through a dialogue, a two-way reaction, between self and society, inside and outside: not a one-way process of either invention-proclamation or injunction-obedience. Of course, the terms will be different under different social conditions.

These meanings of masculinity are grounded in the male body: a body that does things (fights, fornicates) or refuses to do them (doesn't handle weapons, becomes celibate, won't achieve erection). Attention to the linguistic dimension of masculinity leads us (back, it might seem) to the body, both as a source of determining metaphors and as the organizer of human experience. In gender, body and discourse come together in a particularly striking way, as people make meanings using language that ultimately, via many transformations, refers to the sexed body. Body (fleshly, raw and tangible) and mind (abstract, intellectual) are not so far apart as they may tempt us to think. Quarrels involve words as often as blows, particularly those past quarrels that are accessible to us through historical evidence. We could say that gender is culture's imprint of the sexed body in discourse: an imprint which human beings in every society, male and female, constantly test, revise and reshape. They may accept it without difficulty; they may hold it up against themselves to see whether it fits; they may reject it. Still, sometimes in spite of themselves, they cannot do without it. Gender is not the only identity, but there is no identity without gender, even if (as happened more often in the Middle Ages than now) that gender identity involves an apparent refusal of sex.

Knowing about the medieval body often means knowing about the things written about it. Most of our evidence concerning medieval attitudes to the body comes from medical discourse – which set out to describe what bodies were like, or, rather, what they ought to be like. For the history of the Middle Ages, or of any pre-modern moment, an additional touchstone here is the well-known book by Thomas Laqueur, *Making Sex* (1990). Laqueur argued provocatively that before the impact of scientific discoveries between the late seventeenth and early nineteenth centuries, sex followed from gender. The clearer, more important division of humanity was not male from female, but masculine from feminine. Aristotle and his successors defined men and women primarily as to qualities, roles and capabilities, not anatomical difference. So a securely male individual was one who was truly masculine. According to Laqueur, pre-modern anxieties over gender reflect the belief that if social roles did not remain fixed, the body might not either: women might actually become men and vice versa. The consequent fear of gender confusion fits into a context of general anxiety about social change and mobility in early modern culture, beginning in the sixteenth century. Medieval evidence of this exact logic is more difficult to find, which may explain Laqueur's relative lack of attention to the entire era.[4]

Rather than a belief in the subordination of sex to gender, Torke's example shows us an analogue to Laqueur's theory at a symbolic or metaphorical level. Torke's impressive sex organ existed, and exists, in a discourse. It served a purpose in symbolizing certain values which society sustained homosocially. Its power to do so depended, in turn, on its reference to the male body and, especially, Torke's unusual male body. Surely the most famous symbolic structure referring to the male body is the phallus. Like 'masculinity', the word 'phallus' can mean different things. Sometimes it is a euphemistic synonym for the actual, bodily penis. More often, it refers to images of the erect penis, beginning with literal sculptural models venerated in some cultures. Sigmund Freud, the creator of psychoanalysis, recognized that the erect penis has many associations in human language and thought, some overtly sexual and some not: penetrating power, wounding aggression, sexual excitement and male fertility. In the mid-twentieth century, the very influential writings of the French psychoanalyst, Jacques Lacan, focused attention on the meanings of the phallus for power: a fantasy of complete domination and control, which both men and women may seek to possess but which, as a fantasy, is by definition always incompletely achievable. This is the meaning of the phallus that has guided much Lacanian criticism in the humanities. While an overview of Lacanian theory is beyond the scope of this chapter, it is relevant because of its concern with language. Drawing on and adapting Freud's ideas about psychosexual development, Lacan elaborated a model that sees the entry into language and the 'symbolic order' as the originary human trauma. For Lacan, language and abstract thought and, by extension, rationality, intellectuality and the desires for them, are inescapably connected with the masculine: with law, surveillance and control.

So the concept 'phallus' can take us far from tangible human bodies. Is Torke's pintle, revealed to us in discourse, a penis or a phallus? In effect, it is both, because the fantasized, speculative representation of it in the minds of Paston and other men involves both presence and function. Yet other phalluses exist in this story, too. The existence of Torke's children proves his achievement of conventional penetrative sexual capacity, a normal penis. However, their health and size prove something else. In Paston's account, Torke's 'three or four' sons, those big strapping fellows, become metonymically Torke's own phallus, as he wields them

in conversation with the king as an extension of himself, to prove his own sufficient and equivalent manhood. Penetration and domination do not seem as
important as social viability, which Paston then reads backwards into fleshly
sufficiency: phallus becomes penis.

I embarked on this digression about phallicism mainly to show how important
it is, for a historical analysis to be rounded and sensitive, to keep the image of the
real, fleshly body in mind, just as surely as the determining reality of discourse.
Perhaps the historical body we can know is a discursive one, inseparable from all
the discursive strategies that shape it in our written evidence. Yet ultimately it
refers to a 'real' bodily experience. This is what I meant when I described gender
as an identity issue. A full historical account which credits people with full
humanity needs to keep in mind at least how the individual shapes his or her own
perspective, the variety of possible responses there can be. Often, a contradictory
oscillation means that conformity can be as much needed, desired, as imposed,
and can wrestle with an equally strong desire not to conform. Medieval sources
of social history do not always help us to remember these things. And the relevance of gender is not always so obvious or so close to the surface.

Male bodies in society: maturation, competition and self-command

The body receives its cultural apparatus by speaking to society. While the important questions of historical bodily experience (what, for example, did having a
body like Torke's mean for a man's private and social self?) pose severe problems
of evidence, there were many ways the male body fashioned pre-modern masculinity. Genital status and literal phallic symbolism were only the most obvious.
Many apparently 'social' features of masculinity themselves derive ultimately from
the body. Social masculinity presents an irony: men's intensely public lives, as historically recorded, and a discourse that opposed cerebral masculinity to carnal
femininity, distract us from the corporeal nature of masculine signification.

We see this particularly in the evolving, progressive, stage-structured nature of
male social identity. Maturation was central to the achievement of masculinity at
every level. The rawest testaments to the importance of physical maturity are
usually hidden from us by the late Middle Ages. Still, the related vocabulary is eloquent. In medieval England, the words for 'young or immature male' and 'untrust-
worthy male', 'male of no account', blurred easily together. 'Boy' is not to be
understood as meaning 'male child' before the fifteenth century, and its original
meaning of 'servant' or 'person in a menial position' shifted into a term of increasingly vague abuse – but one applied only to males. In fact, it is hard to find the
word 'boy' in a non-derogatory or non-condescending sense before the sixteenth
century. Also 'knave', a common word to designate male children, had a clearly

insulting quality when applied to adult men; its dismissive and diminishing connotations were apparent in the Middle Ages.[5] (The relation of the concept 'woman' to 'girl' or 'maid' was quite unlike that between 'man' and 'boy'. Even where 'maid' implies service, it is not derogatory, and of course where it denotes virginity, it is even less so.)

In a deferential society where maleness conferred certain privileges, and adulthood still others, immaturity could hardly be the source of any positive meanings, especially for males. These linguistic features remind us again that life stages defined men and women in different ways. For the most part, unlike some less complex societies, late medieval Europe had few clear rites of passage that marked the transition from boy to man. Achieving manhood was more a matter of negotiating increasingly competitive masculine hierarchies. Most young men would encounter this feature of social life, regardless of social status, but it was most important, possibly, for the broad middle stratum between the landed elite and the landless destitute. For most of the male population falling between those two extremes, earning a living meant holding one's place in a well-marked order, through an often uneasy balance between deference and self-assertion. John Southwell, servant of a London draper, was trying to achieve this balance when he petitioned Chancery around 1475. John's master, the petition tells us, 'took great displeasure' against John, commanding the young servant to crop the hair off his head. Whether this meant a total shaving or simply a haircut, the point at issue is a subordinate male's control over his own body. John thought 'that he was as other servants be and ought for to be', and refused to shear his locks, upon which, he says, his master convinced a London alderman to authorize John's arrest and imprisonment. This strange quarrel underlines, in an extreme way, the everpresent tension between individual identity – the sense of being an agential person (which, as we have seen, defined masculinity in a prevalent discourse) – and the claims of superior, masterly or lordly authority against that. John attributed his master's actions to a 'malicious appetite': arbitrary, inscrutable will.[6] Of course, female servants, of which many existed, faced the will of masters too, often in equally blunt or brutal ways. Yet playing the competitive game was never as important for feminine identity. Women did not generally become part of a trade hierarchy and while they did negotiate contracts, servanthood was a temporary phase, not part of a career path.

A little Middle English romance named *Gamelyn* shows how literature could image one masculine social code, with very old origins, at a more elite level, and yoke it to physical masculinity. Consisting mostly of bitter, violent yet somewhat comic conflicts between brothers, the story revolves around the struggle of a young elite man to establish himself fully – to embody himself with property. A youngest son cheated out of his inheritance by his false and conniving older brother, Gamelyn regains it essentially through brute strength and full-on assault, turning the tables on his opponents. Gamelyn is not a super-strong man of mythical power, though he does win a wrestling contest and take home a ram and a ring as a prize. His strength enables him to beat up his opponents in a fair (or even unfair) fight. The poem allows physical strength to symbolize a complete kind of competitive masculinity.[7]

The tale of Gamelyn ignores men's relations with women altogether. Yet the hierarchies of household and workplace crucially affected the heterosexual element in gender identity formation at precisely the fraught adolescent period when most young men and women were servants and apprentices. The irony is sharpest in the period after the Black Death, when changed social conditions meant that more young people than ever before took on jobs away from their natal household and community, and arguably had more opportunities to meet and form relationships with others. Nevertheless, apprentices and servants, as John Southwell's petition shows, were in a legal position comparable to children. Their masters had considerable power over them, which included the power to forbid sexual relationships and control marriages. For these young men, others imposed self-command over the body. For some others, it was more facultative but no less obligatory. The books of courtesy that survive, beginning in the early fifteenth century, instruct boys in deportment concerning homosocial contexts. They say nothing meaningful (beyond occasional platitudes) about relations with women, how to speak to women or how to behave in their presence. The imagined requirements are those of men together, and the instructions focus on the correct bearing toward social superiors and attention to speech, gesture, movement and body functions, especially at table.

The identification of sober self-control with masculinity is ancient and was preserved and transmitted to the Middle Ages by the Church, through the traditions of asceticism and monasticism. For clerics of all kinds, celibacy – a very specific form of self-command – was mandatory and definitive. It defines them even more clearly in the eyes of modern historians. With all that we know about the centrality of sexuality in the human psyche, and its particular visibility in masculine discourse, we may wonder about this class of men who were forbidden to have sex or to marry. (This question, by definition, restricts our discussion to the period after about 1100, when clerical celibacy began to be more rigorously enforced. Late antique and early medieval clerics did marry and reproduce.) Apart from celibacy, there were other apparent oddnesses in their social lives; their work did not involve them in ordinary masculine activities and social relations. They must, we imagine, have had difficulty maintaining a masculine gender identity, especially in relating to laymen. The question becomes: did people in medieval society view the celibate clergy as unmanly, effeminate or incomplete because they did not live as laymen did: did not copulate, marry, have children, fight for sport, carry weapons? Did that mean that clerics' private gender identities were also in doubt: that they struggled with the knowledge that they were men in male bodies, with male desires?

Although recent scholarly opinion has tended to say 'yes' to both questions, I think an accurate answer is a matter of balance and a particularly careful weighing of evidence.[8] First, there are extant writings from churchmen who despised sexual desire and seem to have loathed their own bodies' apparent freedom from conscious control, as manifested in spontaneous erections and emissions – men who longed to be free from what they saw as the yoke of carnal temptation. These eloquent testaments come usually from learned clerics, who were often men of influence themselves. A more broadly based class of evidence comes from the abundant discourse on clerical sexual misbehaviour: the Church's injunctions against it, discussions among churchmen on how to prevent or deal with it and proceedings in the ever-convenient ecclesiastical courts against those who had fallen into it. Still, there are some problematic assumptions built into the question. The idea that men are especially sexually needy is a modern one; medieval wisdom held that women were the sex ruled by carnality. So, one can argue, a persistent need for sexual indulgence might itself point to an unmanly lack of self-control. (Sexual behaviour reflected on masculinity mostly through proportion; moderate sexual activity was considered healthy and indeed necessary for men in medical discourse, however much denigrated by the Church.) Whether the lack of demonstrable heterosexuality might lead laymen to suspect homosexual tendencies in clerics (a topos of our own day) is very difficult to establish. Contemporary indications of such suspicions come mostly from satirical poems and stories aimed at the regular clergy, especially the friars, and form part of a larger critique of their alleged hypocrisy and greed. (This reminds us that 'the clergy' was not a monolith.) The more likely stereotype, to judge from both legal and fictional sources, was the heterosexually licentious priest whose close contact with women made him an accomplished seducer.

Clerics fit into a complex world of social relations and the discourse around clerical sexuality was too multivalent, enfolding too many contradictory messages, for it to be reduced to a simple matter of 'less sex equals less manly'. Such a one-dimensional analysis also misses all the other ways in which clerics – including by the late Middle Ages, both regulars and seculars – managed the same gender roles as laymen and interacted socially with them. Like laymen, clerics had to compete for jobs and maintain their place in hierarchies, and the same requirements for deference, patronage and public face-saving pressured their everyday lives like anyone else's. They might be heads of households, consisting of servants and/or other clerics; they might have a bit of land that needed working. They sued and were sued by laymen and other clerics, squabbling not just over moral issues but money and property. Intelligent and lucky young men from the monastic houses could go to university. Parish priests socialized at the local tavern and some of them played football with their parishioners. The relative invisibility of the modern clergy, especially in the Anglo-American world since the 1960s, has desensitized us to these normal nuances of everyday life.

Clerics' enmeshment in the very secular and very masculine world of competition and conflict is demonstrable from a variety of evidence, but for the present purpose I have selected another Chancery petition that is not as colourful or as dramatic as other legal records. Robert Godard, clerk parson of St Olaf's Church in London, petitioned Chancery in 1500 or 1501 on the grounds of his unjust imprisonment and harassment by 'certain evil-disposed persons'. The modern eye skips through the first half of the text, which sounds like a dry and legalistic preamble, to what first appears to be the central point: a 'plaint of rape' brought by an unnamed 'singlewoman' against Godard at the Counter of London. This is worth some attention, because we know that the clergy were sensitive to accusations of sexual misbehaviour and that these could be used against them for several reasons.

Yet the woman's charge operates here only within the context of a masculine dispute. I say masculine, although no one has named Godard's opponents and specified their sex. This assumption is justified not only because of the high statistical probability that any disputants in a medieval legal context would be male. Rather, masculinity relates to the degree to which the problem depends on the regulated function of public institutions staffed and directed exclusively by males – institutions whose intricacies rested on discourses (such as law) generated by and accessible only to men. Godard complains (like John the draper's servant) that others have misrepresented him to the Aldermen of London; that they have effected his imprisonment; that 'in truth they neither had nor have cause of action'; that he has offered 'sufficient sureties, which to take the Sheriffs of London then and there refused' – all of which situates the parson within a thoroughly secular contest: who knows better how to (ab)use the law? Only after he explains all this do we hear that the priest's enemies have 'caused a singlewoman to come to the Counter' and enter her charge of rape.[9] The preamble is not really a preamble at all, but the centre of the case. The woman becomes simply an extension of one masculine response, part of a legal strategy. The priest's social masculinity, defined in terms of his ability to defend himself against other men, is at stake and it is an identity he cannot do without. Still, the proving ground has little to do with sexuality, despite the surface impression. Certain stereotypes and public perceptions made the allegation of heterosexual incontinence a recognized framework for use in attacks against clerics that might have nothing to do with either their sex lives (if any) or, in fact, their clerical status.

Whatever the importance of sexual activity per se, apart from marriage, in the informal group ethic of males, that dimension is largely hidden to us now. When a man married, in contrast, he entered a highly formal structure regulated by institutions, where sexuality interacted with concerns and meanings around authority and property. Body and discourse were never more entwined than here. Right regulation, or governance, was the essence of this form of masculinity, and the household was its setting.

Husbandry: marriage, household and property

In the Middle Ages, and for long after, marriage cemented a man's social identity as certainly as it secured and made respectable a woman's. Marriage meant that a man became the head of a household, and so it granted an authority denied to subordinate men, such as servants and apprentices. It made his children, product of his sexual relations, legitimate, and secured the transmission of his property. Medieval law and custom, of course, granted husbands sweeping powers over women in marriage, which extended to the sexual sphere. Although the Church taught that sexual fidelity (and mutual sexual attention) was equally important for both partners, in practice adulterous wives suffered more severe consequences than adulterous husbands. Some evidence suggests that being named 'cuckold' was worse than 'adulterer' for a man, even in England, where cuckoldry anxiety was not as culturally powerful as in Italy or Spain. One Yorkshire man allegedly called another 'thou whey-bedded cuckold', vividly invoking a marital bed, symbol of household integrity, stained and corrupted by the whey-like semen of another man.[10]

Nevertheless, this does not mean that husbands enjoyed unbridled licence, formal or informal, to indulge their sexuality and aggressivity. Both law and society, for example, easily countenanced the corporal punishment of wives, at a time when children and servants could also expect to be physically chastised. Yet where beatings were clearly excessive or life-threatening, going beyond the bounds of reasonable correction, or indulged in sadistically, the Church might grant a judicial separation and peers and neighbours would express stereotyped disapproval. Sexuality encompasses far more than sexual behaviour. Within pre-modern marriage, where control of property and resources was so heavily weighted in favour of males, this was especially true. So English culture regarded adultery in men differently in terms of severity to that of women. It was in some ways substantially different because of what was at stake and because of the way men could relate to each other through public discourses. A London woman named Elizabeth Montagu sent a rather pathetic petition to the Chancellor in the late fourteenth century, complaining that her husband Thomas had so wasted and misspent his goods, during his adultery with a woman named Margaret, that Elizabeth herself was in danger of being imprisoned for debt.[11] In this discursive strategy, Elizabeth drew on a metaphorical dimension of masculinity that had subtle and pervasive cultural power. The male body of Thomas, unproductively wasting its sexual energies in adultery for the sake of personal indulgence, becomes by implication identified with the property, the estate, which he is also frittering away and wasting. Thomas's private physical body and his public symbolic body, the body that cuts a figure for him in the social world, have become one.

As the wronged wife, Elizabeth can construe this adultery as an offence against her, since it puts her in material danger. Yet English society more often understood male adultery as an offence against other men, because in a sense it was a crime against property: the adulterous male overstepped the limits of his own household and property to encroach on others. This may explain why accusations of male adultery or promiscuity surface in the ecclesiastical courts, where lawyers attempt to discredit their opponents' witnesses, and why they always seem to be accompanied by some form of material or social disrepute (pauperhood, perjury and the like). In contrast, alleged adultery in itself rarely formed the basis of a man's defamation suit. Men's sexual reputations were a secondary factor in their social lives. They could certainly affect a man's standing, but they did not have the central importance evident in the lives of women. Adultery had its primary effect as an indication of insufficient self-command, shown in the command of things outside the self, which might (with other evidence) compromise a man's ability to get along with other men. Clergymen were not immune to this ethic. Sexual misconduct by priests violated both their own vows and the homosocial contract of respect for other men's property; unmanliness in laymen's eyes was just as likely a matter of this twofold betrayal.

An unruled and selfish masculinity was one in which a man abused authority and allowed greed to dominate. Men did not sue over simple accusations of adultery, but they did bring the terms 'poller' and 'extortioner' to court. Polling in this sense means both a shearer of hair and a plunderer. One Norfolk clergyman assailed another in 1523 with, 'Thou art a false polling priest and a shaver, for thou hast used polling, but I will not be polled and shaved on thee'.[12] The meaning of 'extortioner' was broader in the late Middle Ages than now and, along with 'poller', it implies an unjust or oppressive exercise of masculine appetite. John Southwell's master, whose 'malicious appetite' made him command John to 'poll' the hair off his head, was therefore an explicit example. A late medieval verse imagined a time of true justice when 'True men might live without vexation; pollers, promoters, had no domination'.[13] Choosing his words carefully, a London defamation suit witness claimed, 'I called you not extortioner, but I said it is pity you should oppress the poor tenants'.[14] This dimension of masculinity operates at a certain remove from recognizable sexuality, but the body is still present in the background, animating the metaphors. In 1513 Essex, William Giller took exception to being called not only 'false extortioner and poller', but also 'great-bellied whoreson' or, possibly, 'draff bellied churl'.[15] The impact of these insults was likely not restricted to a slight on physical appearance. Great bellies become great for various reasons, but one (certainly in medieval terms) is that someone has filled them with good things, possibly those belonging to others. It is not difficult to see the link either to 'extortioner' or 'poller', since both imply rapacious and unjust greed.

I think the word 'husbandry' serves well to describe this standard of mature and established masculinity. The pun reminds us of the importance of marriage, and the extended meanings show us how a metaphorical male body consisting of property, that is, 'livelihood', needed to be managed and controlled. The stakes were higher for people with more property, of course. Still, the ethic of husbandry was very basic and could operate at any level where marriage made a household.

Masculinity by different measures

I have already made the point that masculinity is a set of meanings sustained in discourse. We now need to consider some more such meanings, and ways of finding them, which are further yet from the obvious masculine roles (husband, father) which surfaced in the last section. This requires yet more attention to language.

Some evidence from defamation suits came up earlier and it is time now to take a closer look. When people defend themselves against perceived injuries to their reputations, they are engaging in a deceptively straightforward manoeuvre that bears directly on their identity. Medieval cases of defamation document this process: reactions to damaging spoken words. The relevant English medieval evidence consists largely of the records of the ecclesiastical courts, the primary forum for most ordinary people to defend their reputations before the Reformation. In England, in the fourteenth and fifteenth centuries, most people suing for defamation in the Church courts were men, and most of them sued other men. Most of the time, the words that initiated the suit involved accusations or insinuations of dishonesty, especially of theft. Fear of the consequences may originally have motivated plaintiffs, should such an accusation be taken seriously by the authorities, but changes in practice by the late fifteenth century meant that general slurs, much vaguer insults, such as 'false thief', with no real danger of the victim's prosecution, could land the speaker in court. In such a context, words only make sense as harmful if they signify a whole set of other associations. And thievery and falseness were the most common slanders litigated by men.[16]

The thief named here, even if literally a thief, is always an umbrella term for a kind of unmasculinity manifested in all trickery, deception and oversubtlety. 'Thief' conveyed stealth and guile, two fundamental vices in English culture. (Consider that English law always treated premeditated and covert crimes more severely.) Chancery petitioners accused their oppressors of 'subtle imagination'.

Guile, cunning and subtle craft were usually unflattering when attributed to anyone, but they had special power to denigrate men. They signalled unmanly deceit, unless there were clear mitigating circumstances. Even the subtlety of law courts may have no place in this system, especially when it has been obviously corrupted. Declared outlaw by the bad brother, who has been made sheriff, Gamelyn simply invades the court set to indict him, physically ejects the justice and convicts and hangs brother, justice and jury. Both Gamelyn and his late-arriving second (good) brother, who stood surety for him, are rewarded in the end by the king; this is true justice.

The unmanly guile signified by the insult 'thief' was intensified by the frequent modifier 'false'. This could go with any insult, but 'false thief' and 'false extortioner' were two frequent ones. Thievery might be implied or stated. A Thaxted man said in 1496, 'Thou art a false thief and a false extortioner, as false as he that hath stolen twenty horses'. 'False' could even stand on its own, as the man knew who said to another, 'ye are a false man to me in your dealing'.[17] It was not a matter purely of implied actions. Falseness and thievery might be inscribed on the body, especially the face. 'Thou false gleand thefe, sayst thou that thou beat not me?' said John Rayner in 1424 Yorkshire, while in Norfolk in 1505, William Saxmundham defended his alleged defamation of Peter Melton thus: 'He said ill to me, he called me gleyed-knave'.[18] Gleand or gleyed means 'side-glanc[ing]', or in modern terms, 'shifty-eyed'. This is the way a thief looks out at the world: planning, watching, assessing. A Kentish man in 1415 told another to take off his hat because he looked like a thief: because his eyes were hooded, one suspects.[19]

These associations lined up against a deep-laid association of masculinity with openness and transparent honesty. The word 'truth', and the idea of true-ness, had an ancient cultural lineage in English that bore especially on masculinity, because of its connection to public discourses deriving ultimately from its oldest meaning of integrity to one's word, honouring the oath: honesty. Indeed, the oath retained great cultural strength even as the growth of written record was undercutting its legal evidentiary power. However, in the late fourteenth century the word began to take on its modern meaning of 'conformity to fact'.[20] In pre-modern England, a 'true man' – a man who was true – was also a 'real man'. A true man was one who honoured his oath. Yet a man who honoured his oath was also a real man because only a man who respected the homosocial ethic enough to deal openly, guilelessly, with other men could establish a masculine social self and be accepted as a real man: a true man in the modern sense. So it is not so surprising that the verse quoted above should oppose 'true men' to pollers and extortioners. Recall the tale of Gamelyn, the young man of exemplary physical strength. His other strength, within the terms of the poem, is moral: he keeps his word and trusts people because he has no 'guile'. This is what enables his brother to fool him in their first conflicts. Even later, Gamelyn gets tricked because his evil brother appeals to his sense of honouring an oath: having rashly 'sworn' that he would bind Gamelyn, the brother asks to bind him so that he not be 'forsworne'.[21] These were durable moral frameworks. Shakespeare exploited the contrast between thief and true man. Much later, Jane Austen had the undeniably masculine Darcy declare, 'Whatever bears affinity to cunning is despicable'. With the important, rule-proving exception of Robin Hood, English culture had not the same fondness for the trickster, the duper of peers, the sneak seducer, as seems evident in the literature of France, Spain or Italy. And there seems little sense that the homosocial competition in England was as intense and bitter as to encourage the kind of closely guarded and jealous masculinity ethic observed in modern resource-poor societies such as Andalusia and Crete.[22] It is not difficult to see how these tendencies shade into recognizable suspicion and disdain for 'imagination' and 'subtlety' in the modern sense: still one of the less savoury features of English culture – and of its exported derivatives.

Yet we do well to remember that ideals and reality often coexist in a contradictory interdependence. In the real world of late medieval England – the insecure, capricious, complicated world – honesty and transparency might need to be compromised by guarded dissimulation, just as prudent husbandry might have to satisfy liberality, or self-control be showily suspended in the performance of raucous self-indulgence. All these dynamics kept the masculine self in focus.

Masculinity is neither a thing, a feature, a system nor an ideology, though at times it resembles these entities in the way people past and present regard and use it. As an identity, it formed a vital part of male persons' self-definition. As a set of meanings, it inhabited the fabric of discourse that linked and animated human subjects and the institutions shaping their lives. So studying medieval masculinity is not only an exercise in the social history of male experience or a reduction of the meanings to one generality such as 'power' or 'dominance'. The meanings, often contradictory, need to be understood as an equilibrium, analogous to the dynamic between interior and exterior features of identity. An analysis that acknowledges body and discourse, sex and gender, is a truly balanced and historical one.

Guide to further reading

Cadden, Joan, *Meanings of Sex Difference in the Middle Ages: Medicine, Science, and Culture* (Cambridge and New York, 1993).

Clover, Carol, 'Regardless of Sex: Men, Women and Power in Early Northern Europe', in Nancy F. Partner (ed.), *Studying Medieval Women: Sex, Gender, Feminism* (Cambridge, MA, 1993), pp. 61–86.

Cohen, Jeffrey Jerome, *Of Giants: Sex, Monsters, and the Middle Ages* (Minneapolis, MN, 1999).

Goldberg, P.J.P. 'Masters and Men in Later Medieval England', in D.M. Hadley (ed.), *Masculinity in Medieval Europe* (London and New York, 1999), pp. 56–70.

Karras, Ruth Mazo, *From Boys to Men: Formations of Masculinity in Late Medieval Europe* (Philadelphia, 2003).

McSheffrey, Shannon, 'Men and Masculinity in Late Medieval London Civic Culture: Governance, Patriarchy, and Reputation', in Jacqueline Murray (ed.), *Conflicted Identities and Multiple Masculinities: Men in the Medieval West* (New York and London, 1999), pp. 243–67.

Roper, Lyndal, 'Introduction', in *Oedipus and the Devil* (London, 1994), pp. 1–34.

Tosh, John, 'What Should Historians Do with Masculinity? Reflections on Nineteenth-Century Britain', *History Workshop Journal* 38 (1994), pp. 179–202.

Notes

1 B. Fagot, 'Psychosocial Determinants of Early Gender Identity', *Annual Review of Sex Research* VI (1995), pp. 17–23.
2 Norman Davis (ed.), *Paston Letters and Papers of the Fifteenth Century* (Oxford, 1971), 1, p. 415.
3 Judith Yanof, 'Barbie and the Tree of Life: The Multiple Functions of Gender in Development', *Journal of the American Psychoanalytic Association* 48 (2000), p. 1439.
4 Thomas Walter Laqueur, *Making Sex: Body and Gender from the Greeks to Freud* (Cambridge, MA, 1990), pp. 4–14, 25–62 and 122–42.
5 E.J. Dobson, 'The Etymology and Meaning of Boy', *Medium aevum* 9 (1940).
6 Public Record Office (PRO), London (Kew), C1/48/107.
7 Neil Daniel (ed.), *The Tale of Gamelyn* (Ann Arbor, MI, 1970), pp. 74–8.
8 P.H. Cullum, 'Clergy, Masculinity and Transgression in Late Medieval England', and R.N. Swanson, 'Angels Incarnate: Clergy and Masculinity from Gregorian Reform to Reformation', both in D.M. Hadley (ed.), *Masculinity in Medieval Europe* (London and New York, 1999).
9 PRO, C1/240/26.
10 Borthwick Institute of Historical Research, York, D/C. CP. 1524/10.

11 PRO, C1/4/116.
12 Basil Cozens-Hardy and Edward Darley Stone (eds), *Norwich Consistory Court Depositions, 1499–1512 and 1518–1530* (London, 1938), no. 294.
13 *Oxford English Dictionary*, 'Poller', citing 'Bradshaw, *St. Werburge*, I. 2401'.
14 London Metropolitan Archives (LMA), DL/C/206, fo. 202v.
15 LMA, DL/C/206, fo. 299r. v.
16 Derek Neal, 'Suits Make the Man: Masculinity in Two English Law Courts, c.1500', *Canadian Journal of History* 37 (2002), esp. pp. 7–13.
17 London Guildhall Library 9065, fo. 243, quoted in L.R. Poos, *A Rural Society after the Black Death: Essex 1350–1525* (Cambridge, 1991), p. 84.
18 Cozens-Hardy and Stone (eds), *Norwich Depositions*, no. 54, and R.H. Helmholz (ed.), *Select Cases on Defamation to 1600* (London, 1985), p. 6.
19 Canterbury Cathedral Archives, X.10.1, fo. 26r.
20 Richard Firth Green, *A Crisis of Truth: Literature and Law in Ricardian England* (Philadelphia, 1999), pp. 1–40, esp. 29.
21 Daniel (ed.) *The Tale of Gamelyn*, pp. 79–80.
22 Stanley H. Brandes, *Metaphors of Masculinity: Sex and Status in Andalusian Folklore* (Philadelphia, 1980), and Michael Herzfeld, *The Poetics of Manhood: Contest and Identity in a Cretan Mountain Village* (Princeton, NJ, 1985).

Index